# Surface Chemistry

# Surface Chemistry

K.R. Desai

Oxford Book Company
Jaipur ▪ New Delhi

*Published By*

**Oxford** Book Company

**Regd. Off.** "Bony Residency", Gate No. 2
Opp. Tilak Public School, Vishweswariya Nagar,
Gopalpura Road, Jaipur - 302018, India
Ph: 0141 - 2761280, Telefax: 2761381
email - oxfordbookjpr@gmail.com
website - oxfordbookcompany.com

**Delhi Office:** 102, 1st Floor, Satyam House,
4327/3, Ansari Road, Darya Ganj,
New Delhi - 110002, India
Ph: 011-45652440

*Distributed By*

**KEYA INTERNATIONAL PRESS LLP**
E-mail: keyainternationalpressllp@gmail.com

Published in India, 2025



ISBN : 9789355244000



Typeset by Shivangi Computers, Jaipur

Printed and bound in India by Thomson Press Ltd.,
New Delhi

# Preface

In most technological applications metals are used either in a finely divided form or in a massive, polycrystalline form. At the microscopic level most materials, with few exceptions, can be considered as a collection or aggregate of single crystal crystallites. The surface chemistry of the material as a whole is therefore crucially dependent upon the nature and type of surfaces exposed on these crystallites. It is therefore vitally important to independently study different, well-defined surfaces. The most commonly employed technique is to prepare macroscopic single crystals of metals and then to deliberately cut them in a way which exposes a large area of the specific surface of interest.

The surface chemistry provides an outstanding surface analytical facility which includes such modern techniques and instruments as scanning, tunnelling microscopy, atomic force microscopy, 50 nm resolution scanning, auger microprobe, HREELS, RVLEED, small spot monochromatic x-ray photoelectron spectroscopy, SIMS, Fourier transform infrared spectroscopy and micro-Raman and scanning electronic microscopy. The surface chemistry pioneers the development of more powerful techniques, for example, tip-based probes for studies in the area of nanostructures and nonomechanics of surfaces and interfaces. It also seeks to modify the properties of surfaces by chemical vapour deposition of metastable materials such as diamond.

The present book endeavours to bring before the reader that the understanding and exploitation of solid state phenomena depended largely on the ability to accurately describe the microscopic structures of surfaces and interfaces and to define correlations between these structures and physical and chemical phenomena. Covering all basic concepts, theories and techniques on surface chemistry, the book provides latest information and developments in a sound but easy to follow theoretical framework.

**K.R. Desai**

# Contents

# 1

# Structure of Metallic Surfaces

## INTRODUCTION

The Surface Chemistry Branch has an outstanding surface analytical facility which includes such modern techniques and instruments as: scanning tunneling microscopy, atomic force microscopy, 50 nm resolution scanning Auger microprobe, high resolution electron energy loss spectroscopy (HREELS), reverse view low energy electron diffraction (RVLEED), small spot (30 and 150 micron) monochromatic X-ray photoelectron spectroscopy, secondary ion mass spectrometry (SIMS), Fourier Transform Infrared (FTIR) spectroscopy, micro-Raman and scanning electronic microscopy. The complementary nuclear techniques of Rutherford Backscattering, Proton Induced X-ray Emission (PIXIE), and nuclear reaction spectroscopies are readily available in another Branch of NRL. The Branch also has available two X-ray absorption spectrometer and one IR beam lines at the National Synchrotron Light Source. In addition, the Branch pioneers the development of more powerful techniques, for example, tip-based probes for studies in the area of nanostructures and nanomechanics of surfaces and interfaces. Novel concepts for sensitive microsensors for air or waterborne toxic chemicals and position sensitive detectors are being pursued. Another programme seeks to modify the properties of surfaces by chemical vapour deposition of metastable materials such as diamond, c-BN and products of superlattice growth of III-V materials.

## SURFACE STRUCTURE OF METALS

In most technological applications, metals are used either in a finely divided form (supported metal catalysts) or in a massive, polycrystalline form electrodes, mechanical fabrications.

At the microscopic level, most materials, with the notable exception of a few truly amorphous specimens, can be considered as a collection or aggregate of single crystal crystallites. The surface chemistry of the material as a whole is therefore crucially dependent upon the nature and type of surfaces exposed on these crystallites.

It is therefore vitally important that we can independently study different, well-defined surfaces. The most commonly employed technique, is to prepare macroscopic (size ~ cm) single crystals of metals and then to deliberately cut-them in a way which exposes a large area of the specific surface of interest.

Most metals only exist in one bulk structural form - the most common metallic crystal structures being:

| | |
|---|---|
| bcc | Body-centred cubic |
| fcc | face-centred cubic |
| hcp | hexagonal close packed |

For each of these crystal systems, there are in principle an infinite number of possible surfaces which can be exposed. In practice, however, only a limited number of planes (predominantly the so-called "low-index" surfaces) are found to exist in any significant amount and we can concentrate our attention on these surfaces. Furthermore, it is possible to predict the ideal atomic arrangement at a given surface of a particular metal by considering how the bulk structure is intersected by the surface. Firstly, however, we need to look in detail at the bulk crystal structures.

## THE *HCP* AND *FCC* STRUCTURES

The *hcp* and *fcc* structures are closely related: they are both based upon stacking layers of atoms, where the atoms are arranged in a close-packed hexagonal manner within the individual layer.

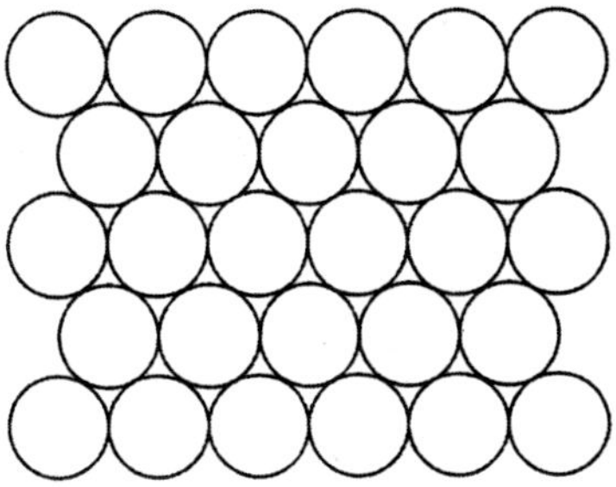

The atoms of the next layer of the structure will preferentially sit in some of the hollows in the first layer - this gives the closest approach of atoms in the two layers and thereby maximizes the cohesive interaction.

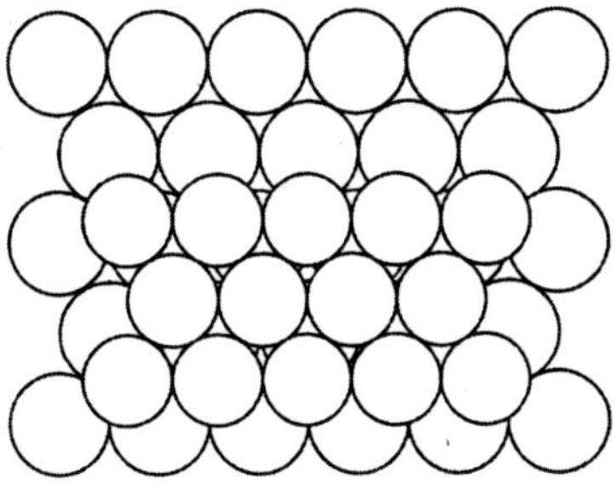

When it comes to deciding where the next layer of atoms should be positioned there are two choices - these differ only in the relative positions of atoms in the 1st and 3rd layers.

In the structure on the left the atoms of the 3rd layer sit directly above those in the 1st layer - this gives rise to the characteristic..ABABA.. packing sequence of the *hcp* structure.

In the structure on the right the atoms of the 3rd layer are laterally offset from those in both the 1st and 2nd layers, and it is not until the 4th layer that the sequence begins to repeat. This is the..ABCABC.. packing sequence of the *fcc* structure. Because of their common origin, both of these structures share common features:

1. The atoms are close packed
2. Each atom has 12 nearest neighbours (CN = 12)

## *FCC* STRUCTURE

Although it is not immediately obvious, the..ABCABC.. packing sequence of the *fcc* structure gives rise to a three-dimensional structure with cubic symmetry.

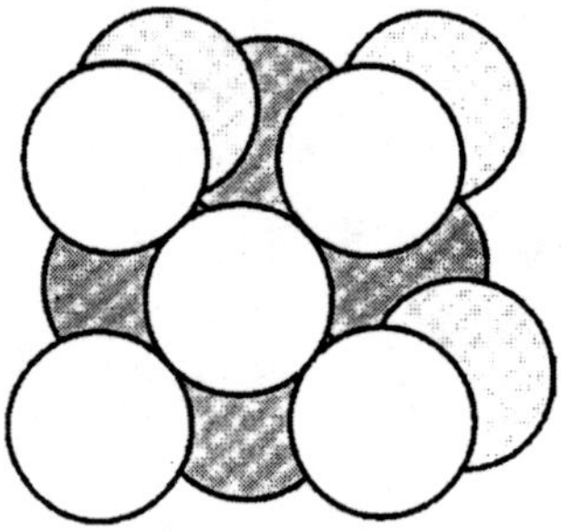

**Fig.** Fcc structure

It is the cubic unit cell that is commonly used to illustrate this structure - but the fact that the origin of the structure lies in the packing of layers of hexagonal symmetry should not be forgotten.

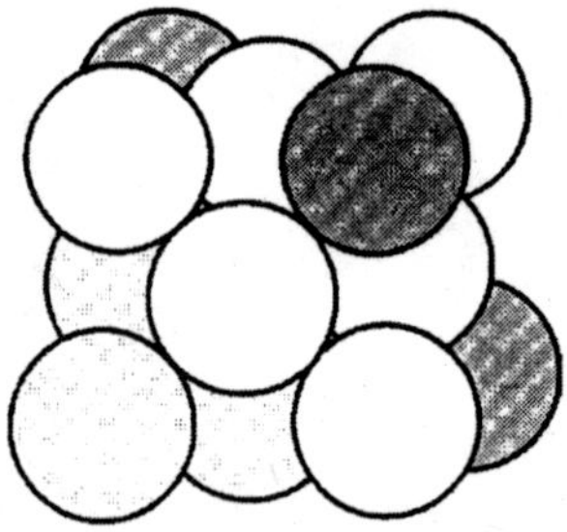

The above diagram shows the atoms of one of the hexagonal close-packed layers highlighted in shades of red, and the atoms of another highlighted in shades of green.

## *HCP* STRUCTURE

The..ABABA.. packing sequence of the *hcp* structure gives rise to a three-dimensional unit cell structure whose symmetry is more immediately related to that of the hexagonally-close packed layers from which it is built, as illustrated in the diagram below.

## THE *BCC* STRUCTURE

The *bcc* structure has very little in common with the *fcc* structure - except the cubic nature of the unit cell. Most importantly, it differs from the *hcp* and *fcc* structures in that it is not a close-packed structure.

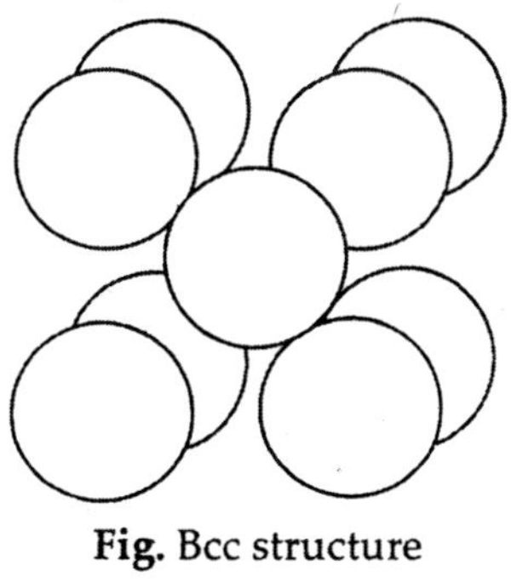

**Fig.** Bcc structure

An ordered surface may be obtained by cutting the three-dimensional bulk structure of a solid along a particular plane to expose the underlying array of atoms. The way in which this plane intersects the three-dimensional structure is very important and is defined by using Miller Indices - this notation is commonly used by both surface scientists and crystallographers since an *ideal* surface of a particular orientation is nothing more than a lattice plane running through the 3D crystal with all the atoms removed from one side of the plane.

In order to see what surface atomic structures are formed on the various Miller index surfaces for each of the different crystal systems we need to consider how the lattice planes bisect the three-dimensional atomic structure of the solid. To pursue this question for the *fcc, hcp* and *bcc* systems you should select the appropriate option from the menu after exiting from this section. As you might expect, however, the various surfaces exhibit a wide range of:

1. Surface symmetry
2. Surface atom coordination and most importantly this results in substantial differences in.
3. Physical properties (electronic characteristics etc.), and
4. Surface chemical reactivity (catalytic activity, oxidation resistance etc.)

## SURFACE STRUCTURE OF FCC METALS

Many of the technologically most important metals possess the *fcc* structure: for example the catalytically important precious metals (Pt, Rh, Pd) all exhibit an *fcc* structure.

The low index faces of this system are the most commonly studied of surfaces: as we shall see they exhibit a range of

1. Surface symmetry
2. Surface atom coordination
3. Surface reactivity

## THE FCC (100) SURFACE

The (100) surface is that obtained by cutting the *fcc* metal parallel to the front surface of the *fcc* cubic unit cell - this exposes a surface (the atoms in blue) with an atomic arrangement of 4-fold symmetry.

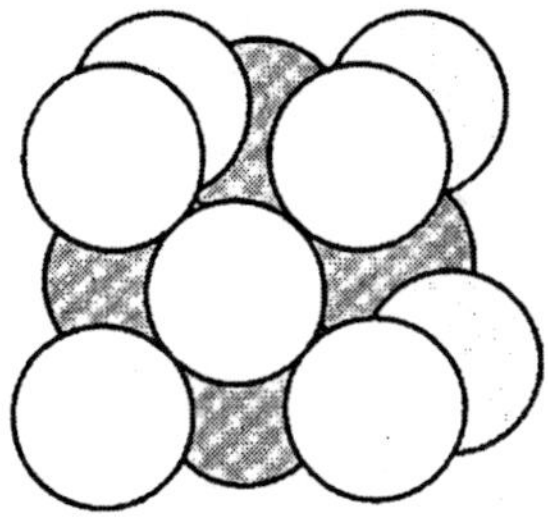

**Fig.** Tcc unit cell (100) face

The diagram below shows the conventional birds-eye view of the (100) surface - this is obtained by rotating the preceding diagram through 45° to give a view which emphasises the 4-fold (rotational) symmetry of the surface layer atoms.

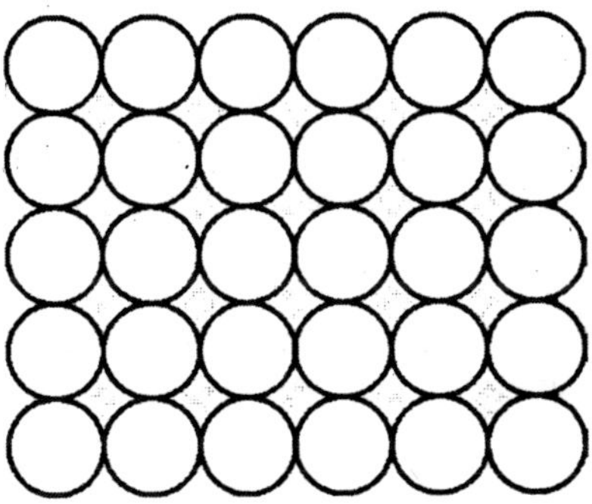

The tops of the second layer atoms are just visible through the holes in the first layer, but would not be accessible to molecules arriving from the gas phase.

There are several other points worthy of note:

1. All the surface atoms are equivalent
2. The surface is relatively smooth at the atomic scale
3. The surface offers various adsorption sites for molecules which have different local symmetries and lead to different coordination geometries - specifically there are:

- On-top sites (above a single metal atom)
- Bridging sites, between two atoms
- Hollow sites, between four atoms

Depending upon the site occupied, an adsorbate species (with a single point of attachments to the surface) is therefore likely to be bonded to either one, two or four metal atoms.

## THE FCC(110) SURFACE

The (110) surface is obtained by cutting the *fcc* unit cell in a manner that intersects the *x* and *y* axes but not the *z*-axis - this exposes a surface with an atomic arrangement of 2-fold symmetry.

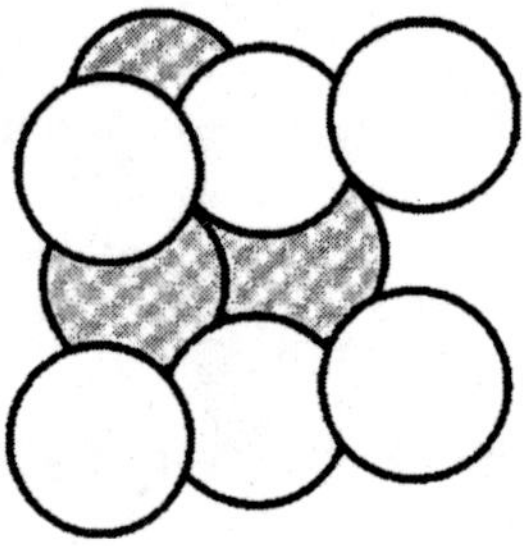

**Fig.** Fcc unit cell (110) face

The diagram below shows the conventional birds-eye view of the (110) surface - emphasising the rectangular symmetry of the surface layer atoms. The diagram has been rotated such that the rows of atoms in the first atomic layer now run vertically, rather than horizontally as in the previous diagram.

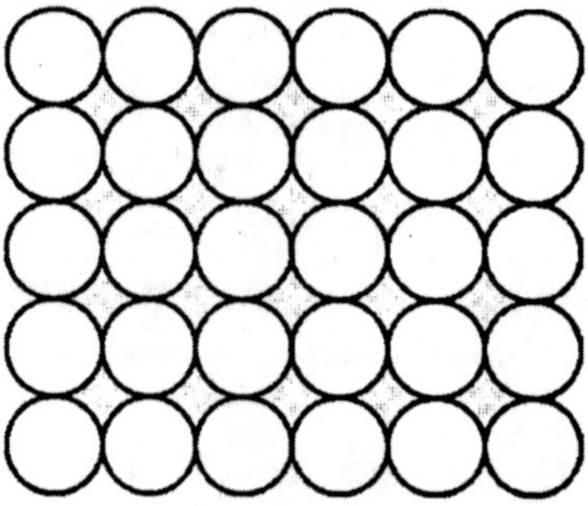

It is clear from this view that the atoms of the topmost layer are much less closely packed than on the (100) surface - in one direction (along the rows) the atoms are in contact i.e. the distance between atoms is equal to twice the metallic (atomic) radius, but in the orthogonal direction there is a substantial gap between the rows.

This means that the atoms in the underlying second layer are also, to some extent, exposed at the surface.

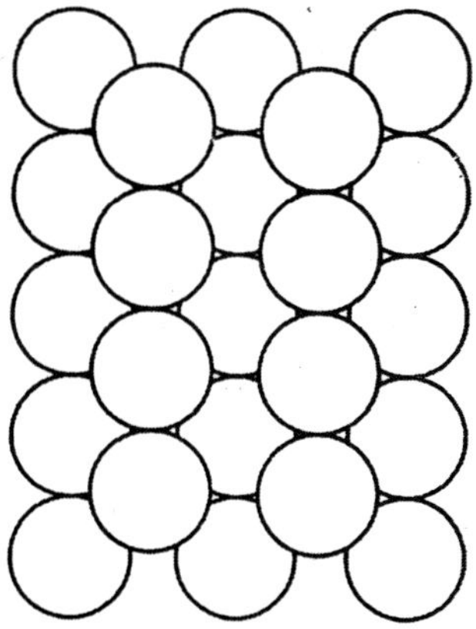

**Fig.** (110) surface plane Cu(110)

The preceding diagram illustrates some of those second layer atoms, exposed at the bottom of the troughs.

In this case, the determination of atomic coordination numbers requires a little more careful thought: one way to double-check your answer is to remember that the CN of atoms in the bulk of the *fcc* structure is 12, and then to subtract those which have been removed from above in forming the surface plane.

If we compare this coordination number with that obtained for the (100) surface, it is worth noting that the surface atoms on a more open ("rougher") surface have a lower CN - this has important implications when it comes to the chemical reactivity of surfaces.

The fact that they are clearly exposed (visible) at the surface implies that they have a lower CN than they would in the bulk.

1. All first layer surface atoms are equivalent, but second layer atoms are also exposed
2. The surface is atomically rough, and highly anisotropic
3. The surface offers a wide variety of possible adsorption sites, including:
   - On-top sites
   - Short bridging sites between two atoms in a single row
   - Long bridging sites between two atoms in adjacent rows
   - Higher coordination sites (in the troughs)

## THE FCC (111) SURFACE

The (111) surface is obtained by cutting the *fcc* metal in such a way that the surface plane intersects the *x*-, *y*- and *z*-axes at the same value - this exposes a surface with an atomic arrangement of 3-fold (apparently 6-fold, hexagonal) symmetry. This layer of surface atoms actually corresponds to one of the close-packed layers on which the *fcc* structure is based.

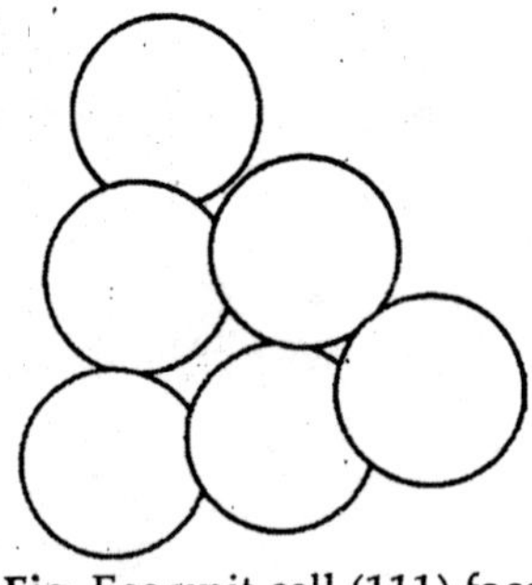

**Fig.** Fcc unit cell (111) face

The diagram below shows the conventional birds-eye view of the (111) surface - emphasising the hexagonal packing of the surface layer atoms. Since this is the most efficient way of packing atoms within a single layer, they are said to be "close-packed".

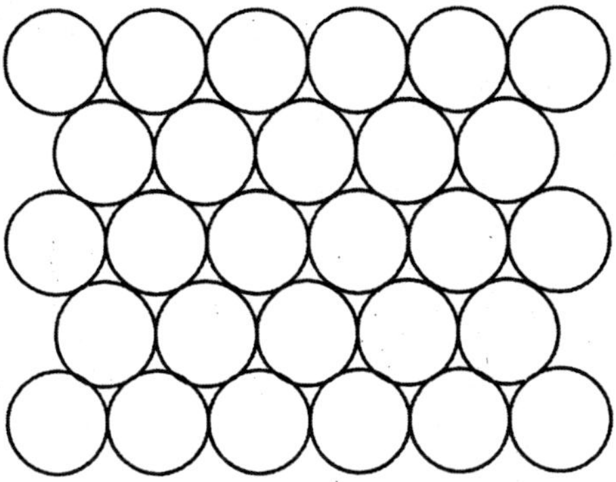

**Fig.** (111) surface plane Pt(111)

The following features are worth noting;

1. All surface atoms are equivalent and have a relatively high CN
2. The surface is almost smooth at the atomic scale
3. The surface offers the following adsorption sites:
   - On-top sites
   - Bridging sites, between two atoms
   - Hollow sites, between three atoms

## HOW DO THESE SURFACES INTERSECT IN IRREGULAR-SHAPED SAMPLES?

Flat surfaces of single crystal samples correspond to a single Miller Index plane and, as we have seen, each individual surface has a well-defined atomic structure. It is these flat surfaces that are used in most surface science investigations, but it is worth a brief aside to consider what type of surfaces exist for an irregular shaped sample (but one that is still based on a single crystal). Such samples can exhibit facets corresponding to a range of different Miller Index planes.

Depending upon how an *fcc* single crystal is cleaved or cut, flat surfaces of macroscopic dimensions which exhibit a wide range of structural characteristics may be produced.

The single crystal surfaces ((100), (110) & (111)) represent only the most frequently studied surface planes of the *fcc* system - however, they are also the most commonly occurring surfaces on such metals and the knowledge gained from studies on this limited selection of surfaces goes a long way in propagating the development of our understanding of the surface chemistry of these metals.

## SURFACE STRUCTURE OF HCP METALS

This important class of metallic structures includes metals such as Co, Zn, Ti & Ru.

The Miller Index notation used to describe the orientation of surface planes for all crystallographic systems is slightly more complex in this case since the crystal structure does not lend itself to description using a standard cartesian set of axes-instead the notation is based upon three axes at 120 degrees in the close-packed plane, and one axis (the *c*-axis) perpendicular to these planes. This leads to a four-digit index structure; however, since the third of these is redundant it is sometimes left out !

## THE HCP (0001) SURFACE

This is the most straightforward of the *hcp* surfaces since it corresponds to a surface plane which intersects only the *c*-axis, being coplanar with the other 3 axes i.e. it corresponds to the close packed planes of hexagonally arranged atoms that form the basis of the structure. It is also sometimes referred to as the (001) surface.

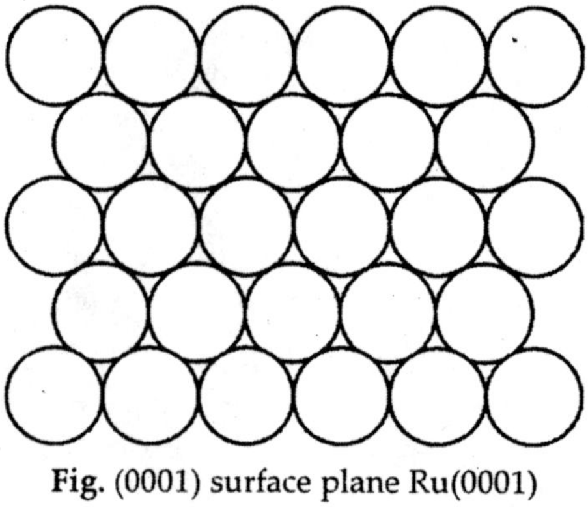

**Fig.** (0001) surface plane Ru(0001)

This conventional plan view of the (0001) surface shows the hexagonal packing of the surface layer atoms.

We can summarise the characteristics of this surface by noting that:

1. All the surface atoms are equivalent and have CN=9
2. The surface is almost smooth at the atomic scale
3. The surface offers the following adsorption sites:
   - On-top sites
   - Bridging sites, between two atoms
   - Hollow sites, between three atoms

## SURFACE STRUCTURE OF BCC METALS

A number of important metals (Fe, W, Mo) have the bcc structure. As a result of the low packing density of the bulk structure, the surfaces also tend to be of a rather open nature with surface atoms often exhibiting rather low coordination numbers.

## THE BCC (100) SURFACE

The (100) surface is obtained by cutting the metal parallel to the front surface of the bcc cubic unit cell - this exposes a relatively open surface with an atomic arrangement of 4-fold symmetry.

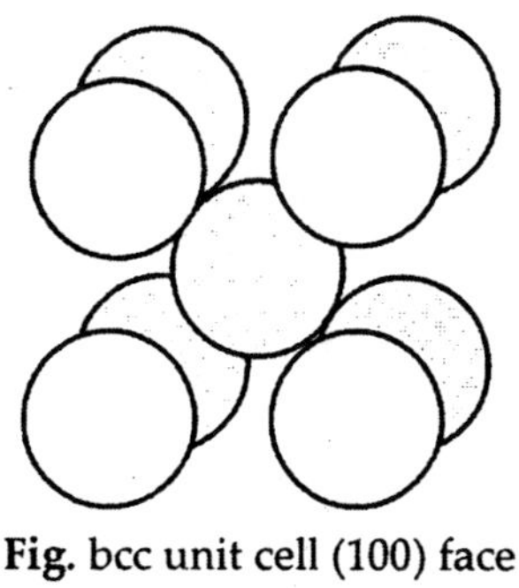

**Fig.** bcc unit cell (100) face

The diagram below shows a plan view of this (100) surface - the atoms of the second layer are clearly visible, although probably inaccessible to any gas phase molecules.

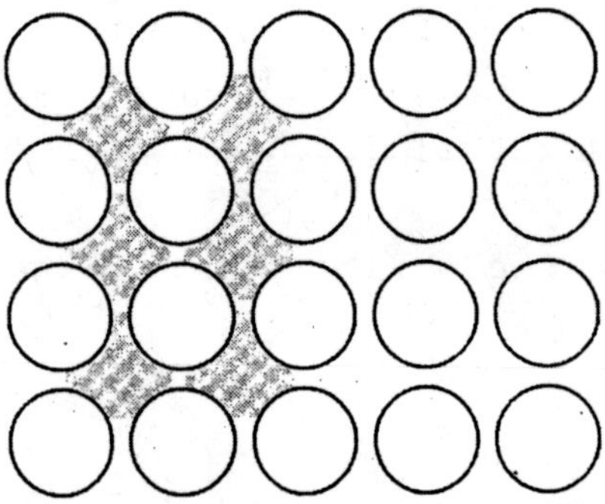

**Fig.** Bcc(100) surface plane Fe(100)

## THE BCC (110) SURFACE

The (110) surface is obtained by cutting the metal in a manner that intersects the x and y axes but creates a surface parallel to the z-axis - this exposes a surface which has a higher atom density than the (100) surface.

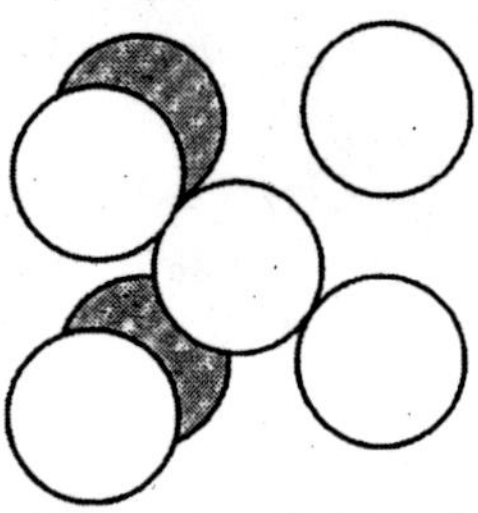

**Fig.** bcc unit cell (110) face

The following diagram shows a plan view of the (110) surface - the atoms in the surface layer strictly form an array of rectangular symmetry, but the surface layer coordination of an individual atom is quite close to hexagonal.

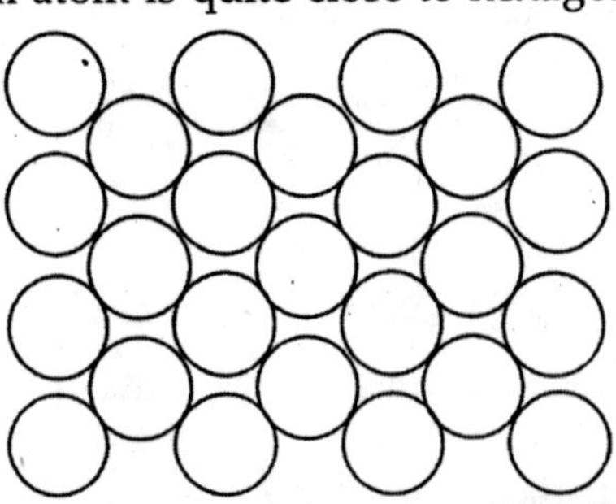

**Fig.** Bcc(110) surface plane Fe(110)

## THE BCC (111) SURFACE

The (111) surface of bcc metals is similar to the (111) face of fcc metals only in that it exhibits a surface atomic arrangement exhibiting 3-fold symmetry - in other respects it is very different.

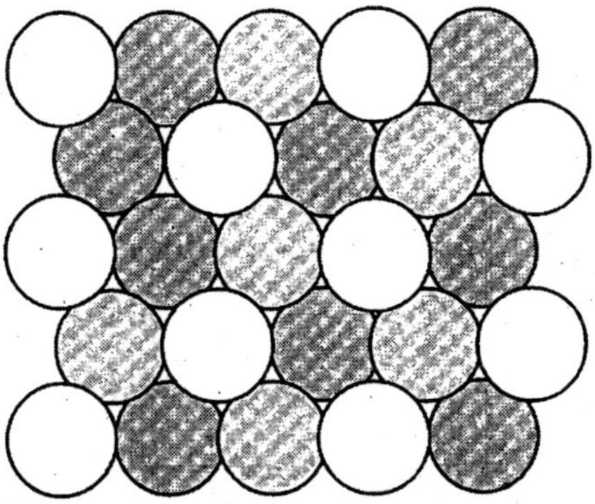

**Fig.** Top View: bcc(111) surface plane Fe(111)

In particular it is a very much more open surface with atoms in both the second and third layers clearly visible when the surface is viewed from above. This open structure is also clearly evident when the surface is viewed in cross-section as shown in the diagram below in which atoms of the various layers have been annotated.

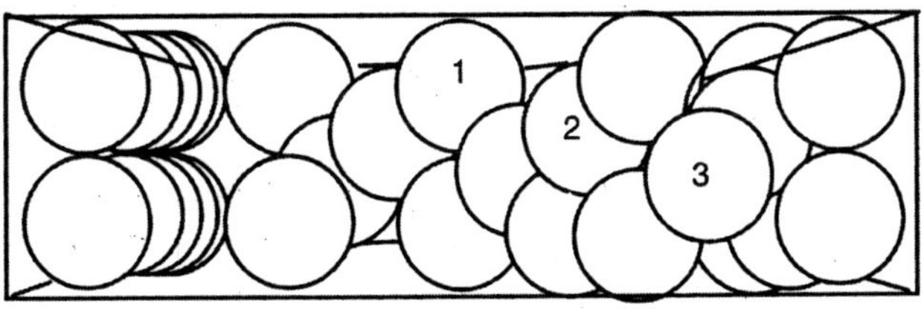


**Fig.** bcc(111) surface plane Fe(111)

## ENERGETICS OF SOLID SURFACES

All surfaces are energetically unfavourable in that they have a positive free energy of formation. A simple rationalisation for why this must be the case comes from considering the formation of new surfaces by cleavage of a solid and recognizing that bonds have to be broken between atoms on either side of the cleavage plane in order to split the solid and create the surfaces. Breaking bonds requires work to be done on the system, so the surface free energy (surface tension) contribution to the total free energy of a system must therefore be positive.

The unfavourable contribution to the total free energy may, however, be minimised in several ways:

1. By reducing the amount of surface area exposed
2. By predominantly exposing surface planes which have a low surface free energy
3. By altering the local surface atomic geometry in a way which reduces the surface free energy

Systems already possessing a high surface energy will not always readily interconvert to a lower energy state at low temperatures due to the kinetic barriers associated with the restructuring - such systems (highly dispersed materials such as those in colloidal suspensions or supported metal catalysts) are thus "metastable".

It should also be noted that there is a direct correspondence between the concepts of "surface stability" and "surface free energy" i.e. surfaces of low surface free energy will be more stable and vice versa.

One rule of thumb, is that the most stable solid surfaces are those with:

1. A high surface atom density
2. Surface atoms of high coordination number

Consequently, for example, if we consider the individual surface planes of an fcc metal, then we would expect the stability to decrease in the order

fcc (111) > fcc (100) > fcc (110)

The presence of a fluid above the surface (gas or liquid) can drastically affect the surface free energies as a result of the possibility of molecular adsorption onto the surface. Preferential adsorption onto one or more of the surface planes can significantly alter the relative stabilities of different planes - the influence of such effects under reactive conditions (the high pressure/high temperature conditions pertaining in heterogeneous catalysis) is poorly understood.

## RELAXATION AND RECONSTRUCTION OF SURFACES

The phenomena of relaxation and reconstruction involve rearrangements of surface (and near surface) atoms, this process being driven by the energetics of the system i.e. the desire to reduce the surface free energy. As with all processes, there may be kinetic limitations which prevent or hinder these rearrangements at low temperatures.

Both processes may occur with clean surfaces in ultrahigh vacuum, but it must be remembered that adsorption of species onto the surface may enhance, alter or even reverse the process!

### Relaxation

Relaxation is a small and subtle rearrangement of the surface layers which may nevertheless be significant energetically, and seems to be commonplace for metal surfaces. It involves adjustments in the layer spacings perpendicular to the surface, there is no change either in the periodicity parallel to the surface or to the symmetry of the surface.

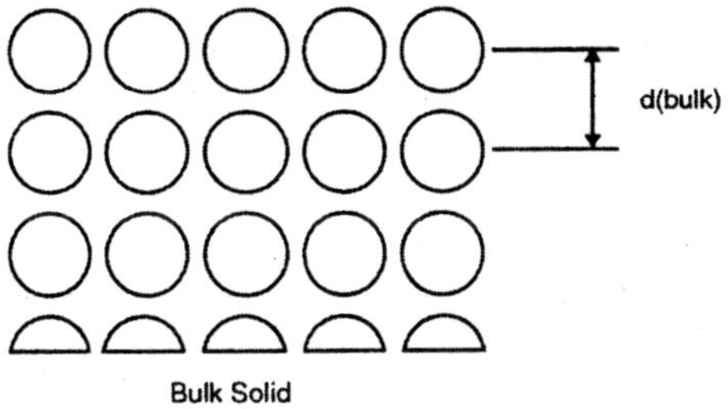


**Fig.** Unrelaxed Surface

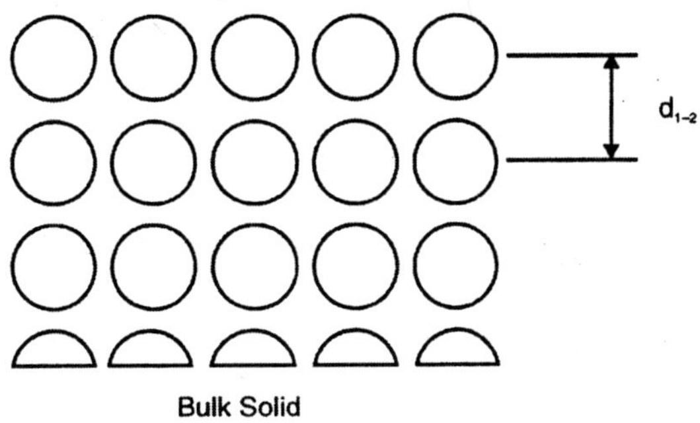


**Fig.** Relaxed Surface ($d_{1-2} < d_{bulk}$)

The lower picture shows the relaxed surface: the first layer of atoms is typically drawn in slightly towards the second layer ($d_{1-2} < d_{bulk}$)

We can consider what might be the driving force for this process at the atomic level....

If we use a localised model for the bonding in the solid then it is clear that an atom in the bulk is acted upon by a balanced, symmetrical set of forces.

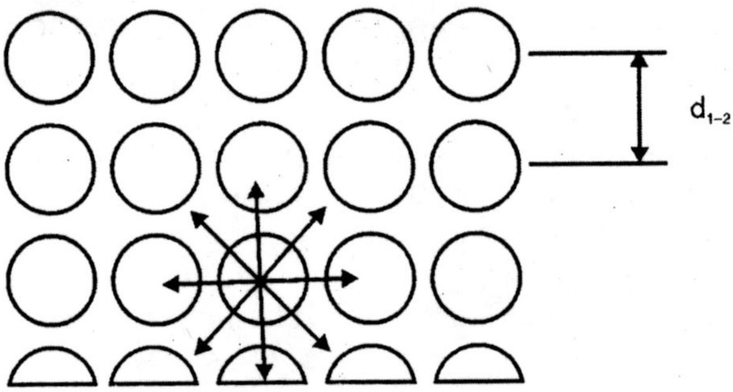


Bulk Solid

On the other hand, an atom at the unrelaxed surface suffers from an imbalance of forces and the surface layer of atoms may therefore be pulled in towards the second layer.

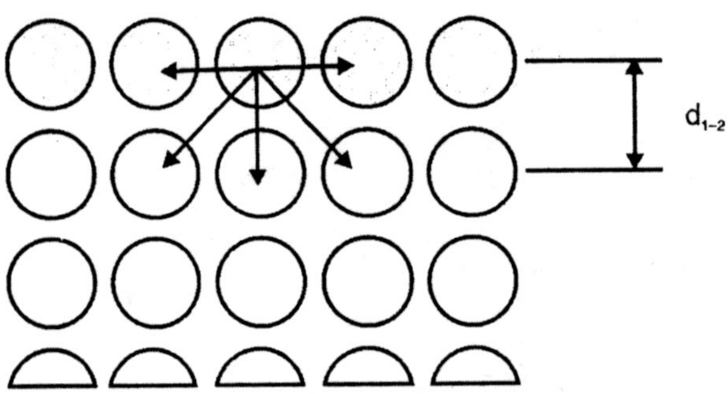


Bulk Solid

The magnitude of the contraction in the first layer spacing is generally small (< 10 %)- compensating adjustments to other layer spacings may extend several layers into the solid.

## Reconstruction

The reconstruction of surfaces is a much more readily observable effect, involving larger (yet still atomic scale) displacements of the surface atoms. It occurs with many of the less stable metal surfaces (e.g. it is frequently observed on fcc(110) surfaces), but is much more prevalent on semiconductor surfaces.

Unlike relaxation, the phenomenon of reconstruction involves a change in the periodicity of the surface structure - the diagram below shows a surface, viewed from the side, which corresponds to an unreconstructed termination of the bulk structure.

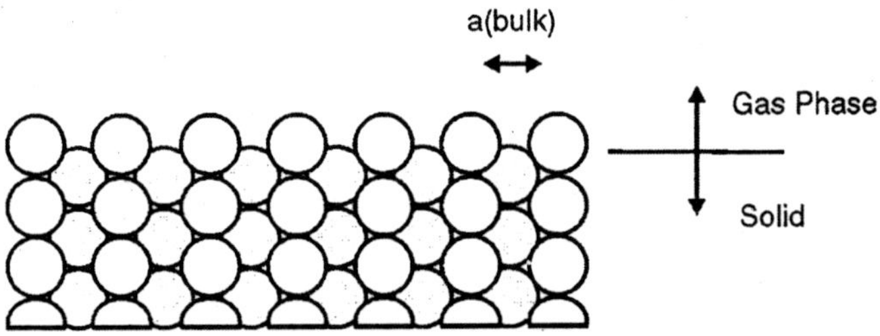


This may be contrasted with the following picture which shows a schematic of a reconstructed surface - this particular example is similar to the "missing row model" proposed for the structure of a number of reconstructed (110) fcc metal surfaces.

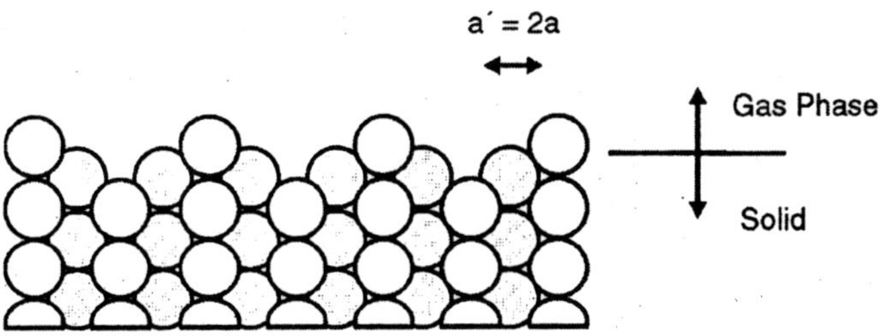


Since reconstruction involves a change in the periodicity of the surface and in some cases also a change in surface symmetry, it is readily detected using surface diffraction techniques.

The overall driving force for reconstruction is once again the minimization of the surface free energy - at the atomic level, however, it is not always clear why the reconstruction should reduce the surface free energy. For some metallic surfaces, it may be that the change in periodicity of the surface induces a splitting in surface-localized bands of energy levels and that this can lead to a lowering of the total electronic energy when the band is initially only partly full.

In the case of many semiconductors, the simple reconstructions can often be explained in terms of a "surface healing" process in which the co-ordinative unsaturation of the surface atoms is reduced by bond formation between adjacent atoms. For example, the formation of a Si(100) surface requires that the bonds between the Si atoms that form the new surface layer and those that were in the layer immediately above in the solid are broken this leaves two "dangling bonds" per surface Si atom.

A relatively small co-ordinated movement of the atoms in the topmost layer can reduce this unsatisfied co-ordination - pairs of Si atoms come together to form surface "Si dimers", leaving only one dangling bond per Si atom. This process leads to a change in the surface periodicity: the period of the surface structure is doubled in one direction giving rise to the so-called (2x1) reconstruction observed on all clean Si(100) surfaces.

Attention has been concentrated on the reconstruction of clean surfaces. It is, however, worth noting that reconstruction of the substrate surface is frequently induced by the adsorption of molecular or atomic species onto the surface - this phenomenon is known as *adsorbate-induced reconstruction* for some examples).

The minimisation of surface energy means that even single crystal surfaces will not exhibit the ideal geometry of atoms to be expected by truncating the bulk structure of the solid parallel to a particular plane. The differences between the real structure of the clean surface and the ideal structure may be imperceptibly small (e.g. a very slight *surface relaxation)* or much more marked and involving a change in the surface periodicity in one or more of the main symmetry directions.

## PARTICULATE METALS

As mentioned in the Introduction, macroscopic single crystals of metals are not generally employed in technological applications.

Massive metallic structures (electrodes etc.) are polycrystalline in nature - the size of individual crystallites being determined by the mechanical treatment and thermal history of the metal concerned. Nevertheless, the nature and properties of the exposed polycrystalline, metal surface is still principally determined by the characteristics of the individual crystal surfaces present. Furthermore, the proportions in which the different crystal surfaces occur is controlled by their relative thermodynamic stabilities. Thus, a macroscopic piece of an fcc metal will generally expose predominantly (111)-type surface planes.

A more interesting case for consideration is that of metals in a highly dispersed system - the classic example of which is a supported metal catalyst. In such catalysts the average metal particle size is invariably sub-micron and may be as small as 1 nm. These metal particles are often tiny single crystals or simple twinned crystals.

The shape of these small crystals is principally determined by the surface free energy contribution to the total energy. There are two ways in which the surface energy can be reduced for a crystal of fixed mass / volume:

1. By minimizing the surface area of the crystallite
2. By ensuring that only surfaces of low surface free energy are exposed.

If matter is regarded as continuous then the optimum shape for minimizing the surface free energy is a sphere this is why liquid droplets in free space are basically spherical.

Unfortunately, we cannot ignore the discrete, atomic nature of matter and the detailed atomic structure of surfaces when considering particles of the size found in catalysts. If, for example, we consider an fcc metal (eg. Pt) and ensure that only the most stable (111)-type surfaces are exposed, then we end up with a crystal which is an octahedron.

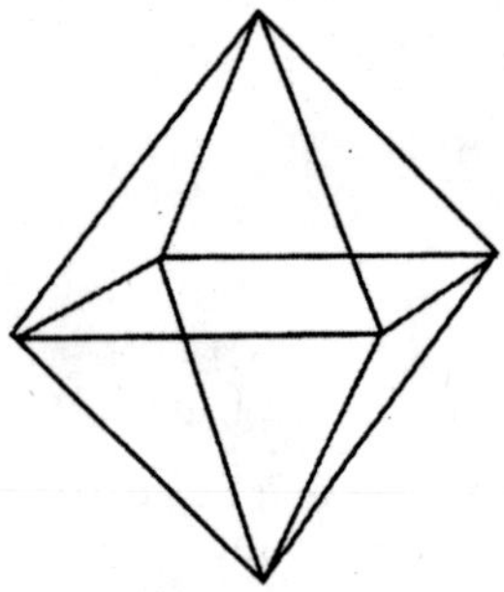

**Fig.** Octahedron exposing 8 symmetry-related fcc(111)-type faces

*Note:* there are 8 different, but crystallographically-equivalent, surface planes which have the (111) surface structure - the {111} faces. They are related by the symmetry elements of the cubic fcc system.

A compromise between exposing only the lowest energy surface planes and minimizing the surface area is obtained by truncating the vertices of the octahedron - this generates a cubo-octahedral particle as shown below, with 8 (111)-type surfaces and 6 smaller, (100)-type surfaces and gives a lower (surface area / volume) ratio.

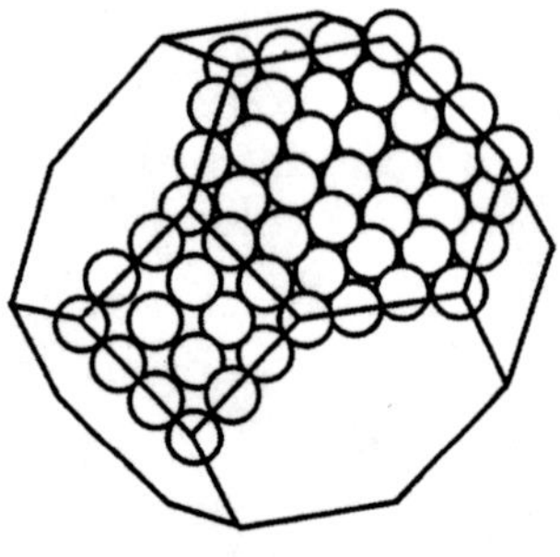

Crystals of this general form are often used as conceptual models for the metal particles in supported catalysts.

The atoms in the middle of the {111} faces show the expected CN=9 characteristic of the (111) surface. Similarly, those atoms in the centre of the {100} surfaces have the characteristic CN=8 of the (100) surface.

However, there are also many atoms at the corners and intersection of surface planes on the particle which show lower coordination numbers.

This model for the structure of catalytic metal crystallites is not always appropriate: it is only reasonable to use it when there is a relatively weak interaction between the metal and the support phase (many silica supported catalysts).

A stronger metal-support interaction is likely to lead to increased "wetting" of the support by the metal, giving rise to:

- A greater metal-support contact area
- A significantly different metal particle morphology

*For example:*

In the case of a strong metal-support interaction the metal/oxide interfacial free energy is low and it is inappropriate to consider the surface free energy of the metal crystallite in isolation from that of the support.

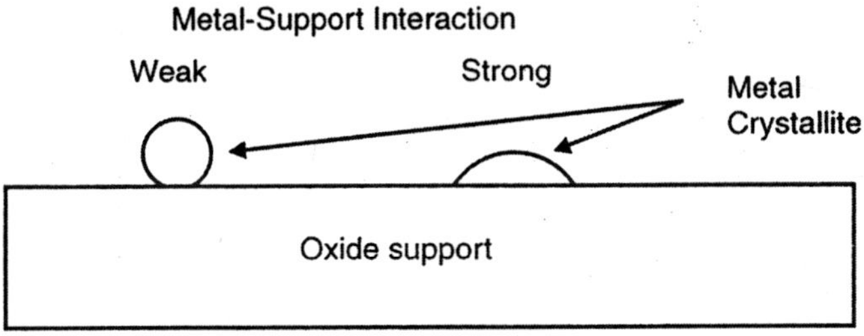


Our knowledge of the structure of very small particles comes largely from high resolution electron microscopy (HREM) studies - with the best modern microscopes it is possible to directly observe such particles and resolve atomic structure.

## OTHER SINGLE CRYSTAL SURFACES

There are many other types of single crystal surface which may be prepared and studied. These include:

- High index surfaces of metals
- Single crystal surfaces of compounds

These will not be covered in any depth, but a few illustrative examples are given below to give you a flavour of the additional complexity involved when considering such surfaces.

## High Index Surfaces of Metals

High index surfaces are those for which one or more of the Miller Indices are relatively large numbers. The most commonly studied surfaces of this type are vicinal surfaces which are cut at a relatively small angle to one of the low index surfaces. The ideal surfaces can then be considered to consist of terraces which have an atomic arrangement identical with the corresponding low index surface, separated by monatomic steps.

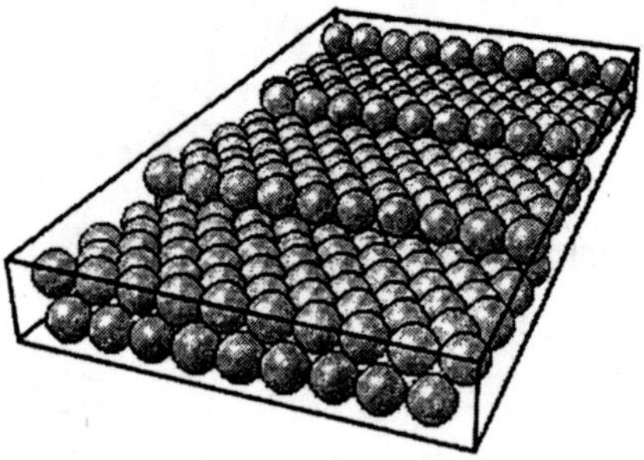

**Fig.** Perspective view of the fcc(775) surface

As seen above, the ideal fcc(775) surface has a regular array of such steps and these steps are both straight and parallel to one another.

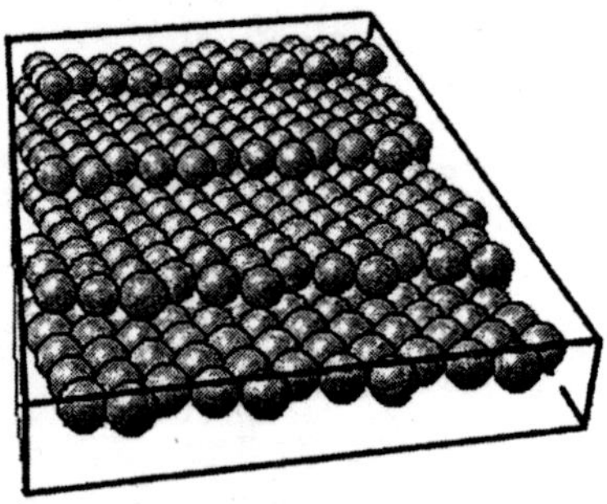

**Fig.** Perspective view of the fcc(10.8.7) surface

By contrast a surface for which all the Miller indices differ must not only exhibit steps but must also contain kinks in the steps. An example of such a surface is the fcc surface the ideal geometry.

Real vicinal surfaces do not, of course, exhibit the completely regular array of steps and kinks illustrated for the ideal surface structures, but they do exhibit this type of step and terrace morphology. The special adsorption sites available adjacent to the steps are widely recognised to have markedly different characteristics to those available on the terraces and may thus have an important role in certain types of catalytic reaction.

## Single Crystal Surfaces of Compounds

The ideal surface structures of the low index planes of compound materials can be easily deduced from the bulk structures in exactly the same way as can be done for the basic metal structures. For example, the NaCl(100) surface that would be expected from the bulk structure is shown below:

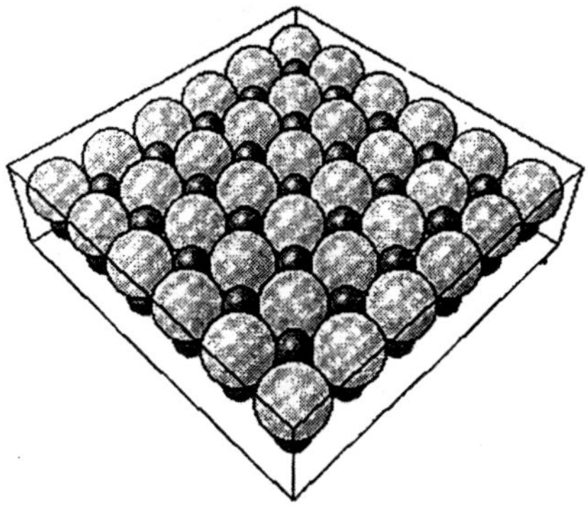

**Fig.** Perspective view of the NaCl(100) surface

In addition to the relaxation and reconstruction exhibited by elemental surfaces, the surfaces of compounds may also show deviations from the bulk stoichiometry due to surface localised reactions and/or surface segregation of one or more components of the material.

For further information on the surface structure of compound materials you should take a look at:

# 2

# Adsorption of Molecules on Surfaces

## INTRODUCTION TO MOLECULAR ADSORPTION

The adsorption of molecules on to a surface is a necessary prerequisite to any surface mediated chemical process.

For example, in the case of a surface catalysed reaction it is possible to break down the whole continuously-cycling process into the following five basic steps:

1. Diffusion of reactants to the active surface
2. Adsorption of one or more reactants onto the surface
3. Surface reaction
4. Desorption of products from the surface
5. Diffusion of products away from the surface

The above scheme not only emphasises the importance of the adsorption process but also its reverse - namely desorption. It is these two processes which are considered in this Section.

## NOTES ON TERMINOLOGY:

*Substrate*: frequently used to describe the solid surface onto which adsorption can occur; the substrate is also occasionally (although not here) referred to as the adsorbent.

*Adsorbate*: the general term for the atomic or molecular species which are adsorbed (or are capable of being adsorbed) onto the substrate.

*Adsorption* - the process in which a molecule becomes adsorbed onto a surface of another phase

*Coverage* - a measure of the extent of adsorption of a species onto a surface.

Usually denoted by the lower case Greek "theta", q

*Exposure* - a measure of the amount of gas which a surface has seen; more specifically, it is the product of the pressure and time of exposure (normal unit is the Langmuir, where 1 L = $10^{-6}$ Torr s).

## HOW DO MOLECULES BOND TO SURFACES ?

There are two principal modes of adsorption of molecules on surfaces:

Physical Adsorption (Physisorption)

Chemical Adsorption (Chemisorption)

The basis of distinction is the nature of the bonding between the molecule and the surface. With...

*Physical Adsorption*: the only bonding is by weak Van der Waals - type forces. There is no significant redistribution of electron density in either the molecule or at the substrate surface.

*Chemisorption*: a chemical bond, involving substantial rearrangement of electron density, is formed between the adsorbate and substrate. The nature of this bond may lie anywhere between the extremes of virtually complete ionic or complete covalent character.

## TYPICAL CHARACTERISTICS OF ADSORPTION PROCESSES

The most definitive method for establishing the formation of a chemical bond between the adsorbing molecule and the substrate (i.e. chemisorption) is to use an appropriate spectroscopic technique, for example

IR to observe the vibrational frequency of the substrate/ adsorbate bond.

| | Chemisorption | Physisorption |
|---|---|---|
| Temperature Range (over which adsorption occurs) | Virtually unlimited (but a given molecule may effectively adsorb only over a small range) | Near or below the condensation point of the gas (e.g. Xe < 100 K, $CO_2$ < 200 K) |
| Adsorption Enthalpy | Wide range (related to the chemical bond strength) - typically 40 - 800 kJ $mol^{-1}$ | Related to factors like molecular mass and polarity but typically 5-40 kJ $mol^{-1}$ (i.e. ~ heat of liquefaction) |
| Crystallographic Specificity (variation between different surface planes of the same crystal) | Marked variation between crystal planes | Virtually independent of surface atomic geometry |
| Nature of Adsorption | Often dissociative<br>May be irreversible | Non-dissociative<br>Reversible |
| Saturation Uptake | Limited to one monolayer | Multilayer uptake possible |
| Kinetics of Adsorption | Very variable - often an activated process | Fast - since it is a non-activated process |

UPS to monitor intensity & energy shifts in the valence orbitals of the adsorbate and substrate.

## ADSORPTION KINETICS - THE RATE OF ADSORPTION

The rate of adsorption, $R_{ads}$, of a molecule onto a surface can be expressed in the same manner as any kinetic process. For example, when it is expressed in terms of the partial pressure of the molecule in the gas phase above the surface:

$$R_{ads} = k' P^x$$

where: $x$ - kinetic order

$k'$ - rate constant

$P$ - partial pressure

If the rate constant is then expressed in an Arrhenius form, then we obtain a kinetic equation of the form:

$$R_{ads} = A \exp(-E_a/RT).P^x$$

where $E_a$ is the *activation energy for adsorption*, and $A$ the *pre-exponential (frequency) factor*.

It is much more informative, however, to consider the factors controlling this process at the molecular level....

The rate of adsorption is governed by

1. The rate of arrival of molecules at the surface
2. The proportion of incident molecules which undergo adsorption

i.e. we can express the rate of adsorption (per unit area of surface) as a product of the incident molecular flux, $F$, and the sticking probability, $S$.

$$R_{ads} = S.\ F \text{ [molecules m}^{-2}\text{s}^{-1}]$$

The flux of incident molecules is given by the Hertz-Knudsen equation

$$\text{Flux, } F = P\ /\ (2\text{p}mkT)^{1/2} \text{ [molecules m}^{-2}\text{s}^{-1}]$$

where

$P$ - gas pressure [N m$^{-2}$]

$m$ - mass of one molecule [kg]

$T$ - temperature [K]

The sticking probability is clearly a property of the adsorbate / substrate system under consideration but must lie in the range $0 < S < 1$; it may depend upon various factors - foremost amongst these being the existing coverage of adsorbed species ($\theta$) and the presence of any activation barrier to adsorption. In general,therefore,

$$S = f(\theta).\ \exp(-E_a/RT)$$

where, once again, $E_a$ is the activation energy for adsorption and $f(\theta)$ is some, as yet undetermined, function of the existing surface coverage of adsorbed species.

Combining the equations for $S$ and $F$ yields the following expression for the rate of adsorption:

$$\text{R} = \frac{f(\theta)\cdot P}{\sqrt{2\pi mkT}} \exp\left(-E_a / RT\right)$$

1. The above equation indicates that the rate of adsorption is expected to be first order with regard to the partial pressure of the molecule in the gas phase above the surface.
2. It should be recognised that the activation energy for adsorption may itself be dependent upon the surface coverage, i.e. $E_a = E(\theta)$.
3. If it is further assumed that the sticking probability is directly proportional to the concentration of vacant surface sites (which would be a reasonable first approximation for non-dissociative adsorption) then $f(\theta)$ is proportional to $(1-\theta)$; where, in this instance, q is the fraction of sites which are occupied (i.e. the Langmuir definition of surface coverage).

The factors which determine the magnitude of the activation energy of adsorption you should which looks at the typical PE curve associated with various types of adsorption process.

## ESTIMATING SURFACE COVERAGES ARISING AS A RESULT OF GAS EXPOSURE

If a surface is initially clean and it is then exposed to a gas pressure under conditions where the rate of desorption is very slow, then the coverage of adsorbed molecules may initially be estimated simply by consideration of the kinetics of adsorption.

As noted above, the rate of adsorption is given by: $R_{ads} = S.F$

i.e.

$$\frac{dN_{ads}}{dt} = SF$$

where: $N_{ads}$ is the number of adsorbed species per unit area of surface.

In general, this equation must be integrated to obtain an expression for $N_{ads}$, since the sticking probability is coverage (and hence also time) dependent.

However, if it is assumed that the sticking probability is essentially constant (which may be a reasonable approximation for relatively low coverages), then this integration simply yields:

$$N_{ads} = SFt$$

## PE CURVES AND ENERGETICS OF ADSORPTION

In this section we will consider both the energetics of adsorption and factors which influence the kinetics of adsorption by looking at the "potential energy diagram/curve" for the adsorption process.

The potential energy curve for the adsorption process is a representation of the variation of the energy (PE or $E$) of the system as a function of the distance ($d$) of an adsorbate from a surface.

Within this simple one-dimensional (1D) model, the only variable is the distance ($d$) of the adsorbing molecule from the substrate surface. Thus, the energy of the system is a function only of this variable i.e. $E = E(d)$

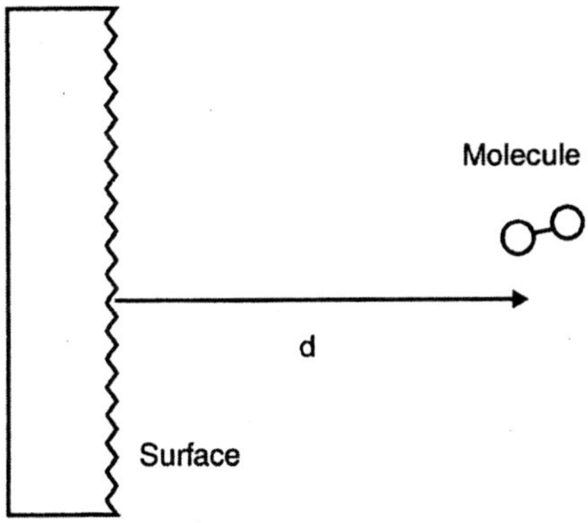


It should be remembered that this is a very simplistic model which neglects many other parameters which influence the energy of the system (a single molecule approaching a clean surface), including for example

- The angular orientation of the molecule
- Changes in the internal bond angles and bond lengths of the molecule

- The position of the molecule parallel to the surface plane

The interaction of a molecule with a given surface will also clearly be dependent upon the presence of any existing adsorbed species, whether these be surface impurities or simply pre-adsorbed molecules of the same type (in the latter case we are starting to consider the effect of surface coverage on the adsorption characteristics).

Nevertheless it is useful to first consider the interaction of an isolated molecule with a clean surface using the simple 1D model. For the purposes of this tutorial we will also not be overly concerned whether the "energy" being referred to should strictly be the internal energy, the enthalpy or free energy of the system.

**Physisorption**

In the case of pure physisorption (e.g. Ar / metals), the only attraction between the adsorbing species and the surface arises from weak, van der Waals forces.

As illustrated below, these forces give rise to a shallow minimum in the PE curve at a relatively large distance from the surface (typically $d > 0.3$ nm) before the strong repulsive forces arising from electron density overlap cause a rapid increase in the total energy.

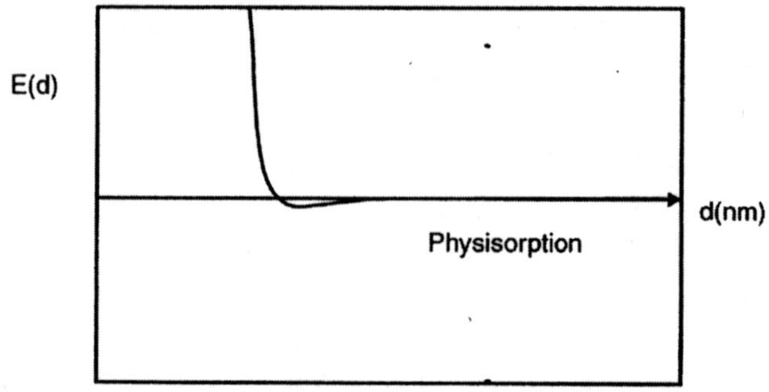


There is no barrier to prevent the atom or molecule which is approaching the surface from entering this physisorption well, i.e. the process is not activated and the kinetics of physisorption are invariably fast.

**Physisorption + Molecular Chemisorption**

The weak physical adsorption forces and associated long-range attraction will be present to varying degrees in all adsorbate / substrate systems. However, in cases where chemical bond formation between the adsorbate and substrate can also occur, the PE curve is dominated by a much deeper chemisorption minimum at shorter values of *d*.

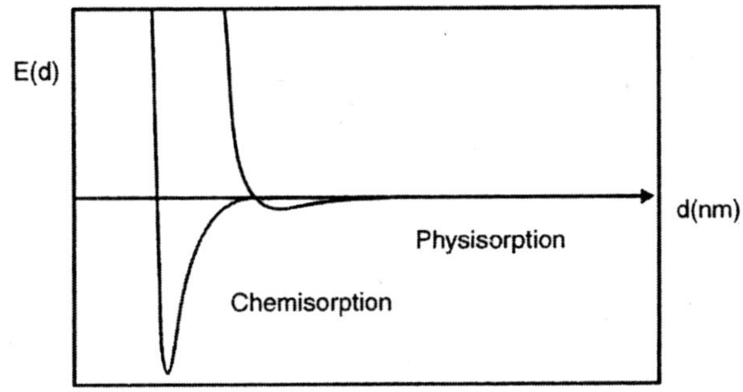


The graph above shows the PE curves due to physisorption and chemisorption separately - in practice, the PE curve for any real molecule capable of undergoing chemisorption is best described by a combination of the two curves, with a curve crossing at the point at which chemisorption forces begin to dominate over those arising from physisorption alone.

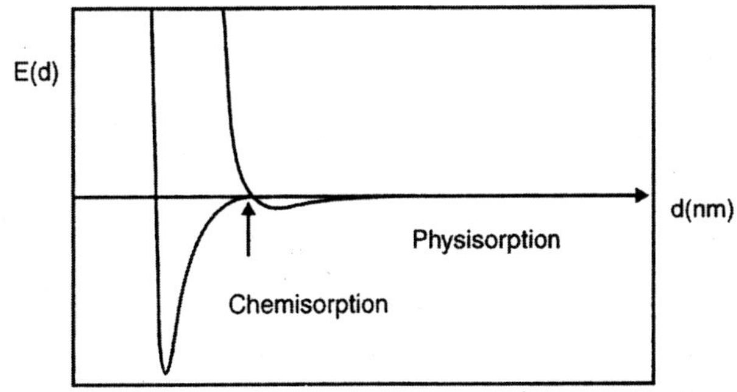


The minimum energy pathway obtained by combining the two PE curves is now highlighted in red. Any perturbation of the combined PE curve from the original, separate curves is most likely to be evident close to the highlighted crossing point.

For clarity, we will now consider only the overall PE curve:

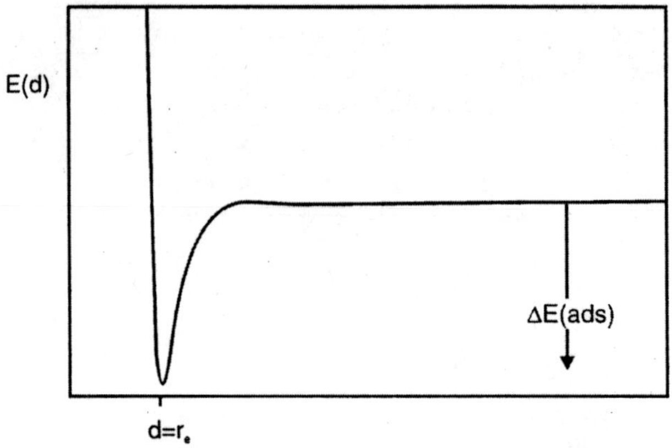


The depth of the chemisorption well is a measure of the strength of binding to the surface - in fact it is a direct representation of the energy of adsorption, whilst the location of the global minimum on the horizontal axis corresponds to the equilibrium bond distance ($r_e$) for the adsorbed molecule on this surface.

The energy of adsorption is *negative,* and since it corresponds to the energy *change* upon adsorption it is better represented as $\Delta E$(ads) or $\Delta E_{ads}$. However, you will also often find the depth of this well associated with the enthalpy of adsorption, $\Delta H$(ads).

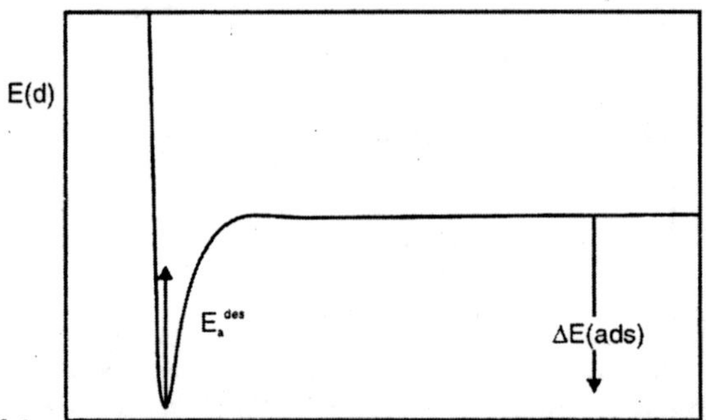


In this particular case, there is clearly no barrier to be overcome in the adsorption process and there is no activation energy of adsorption.

There is of course a significant barrier to the reverse, desorption process the red arrow in the diagram below represents the activation energy for desorption.

Clearly in this particular case, the magnitudes of the energy of adsorption and the activation energy for desorption can also be equated i.e.

$E_a^{des} = \Delta E$ (ads) or $E_a^{des} \sim -\Delta H$ (ads)

**Physisorption + Dissociative Chemisorption**

In this case the main differences arise from the substantial changes in the PE curve for the chemisorption process

Again, we start off with the basic PE curve for the physisorption process which represents how the molecule can weakly interact with the surface:

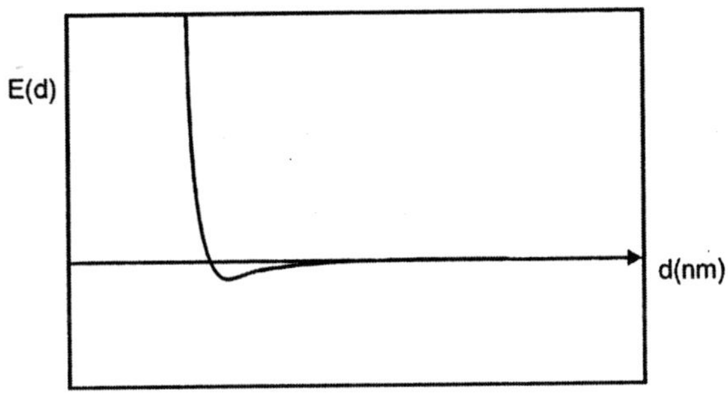


If we now consider a specific molecule such as $H_2$ and initially treat it as being completely isolated from the surface (i.e. when the distance, *d*, is very large) then a substantial amount of energy has to be put into the system in order to cause dissociation of the molecule.

$H_2 \rightarrow H + H$

- this is the bond dissociation energy [*D*(H-H)], some 435 kJ mol$^{-1}$ or 4.5 eV.

The red dot in the diagram above thus represents two hydrogen atoms, equidistant (and a long distance) from the surface and also now well separated from each other. If these atoms are then allowed to approach the surface they may ultimately both form strong chemical bonds to the substrate.... this corresponds to the minimum in the red curve which represents the chemisorption PE curve for the two H atoms.

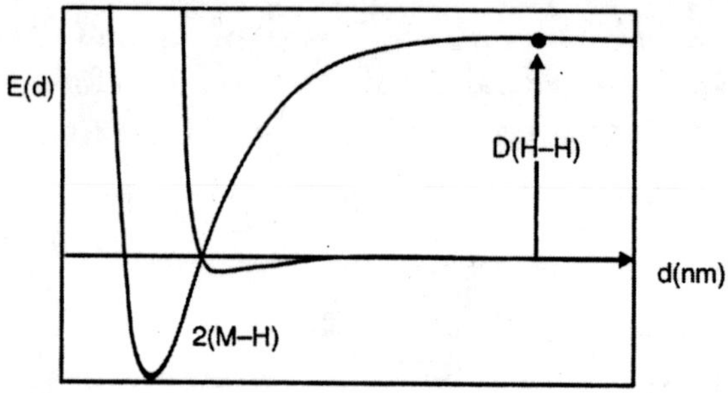


In reality, of course, such a mechanism for dissociative hydrogen chemisorption is not practical - the energy downpayment associated with breaking the H-H bond is far too severe.

Instead, a hydrogen molecule will initially approach the surface along the physisorption curve. If it has sufficient energy it may pass straight into the chemisorption well ("direct chemisorption").

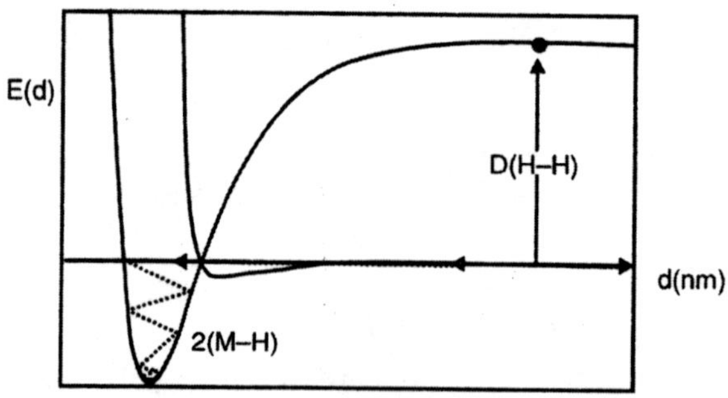


or, alternatively, it may first undergo transient physisorption - a state from which it can then either desorb back as a molecule into the gas phase or cross over the barrier into the dissociated, chemisorptive state (as illustrated *schematically* below).

In this latter case, the molecule can be said to have undergone "precursor-mediated" chemisorption.

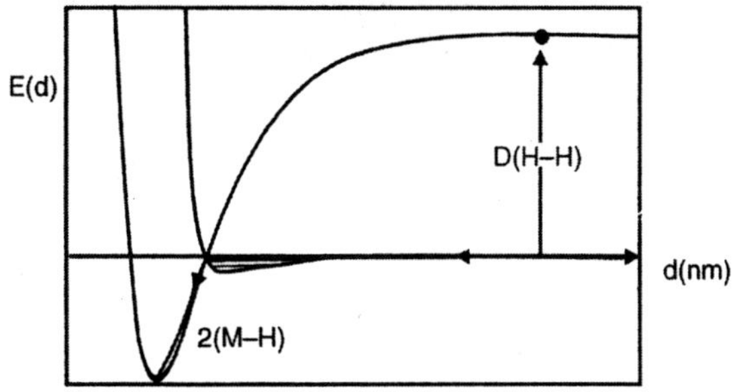


The characteristics of this type of dissociative adsorption process are clearly going to be strongly influenced by the position of the crossing point of the two curves (molecular physisorption v's dissociative chemisorption) - relatively small shifts in the position of either of the two curves can significantly alter the size of any barrier to chemisorption.

In the example immediately below there is no direct activation barrier to dissociative adsorption - the curve crossing is below the initial "zero energy" of the system.

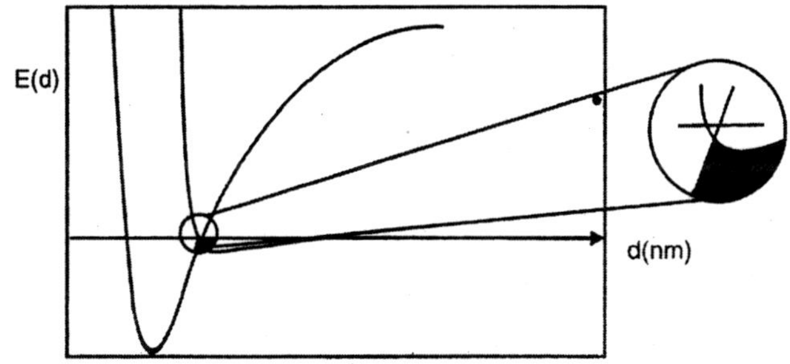


whilst,

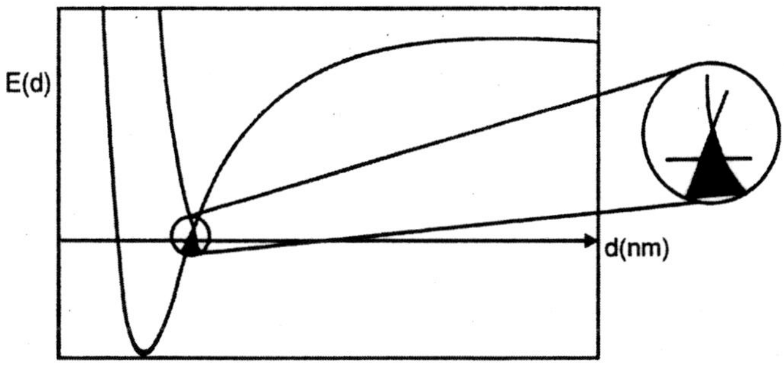

there is a substantial barrier to chemisorption. Such a barrier has a major influence on the kinetics of adsorption.

At this point it is useful to return to consider the effect of such a barrier on the relationship between the activation energies for adsorption and desorption, and the energy (or enthalpy) of adsorption.

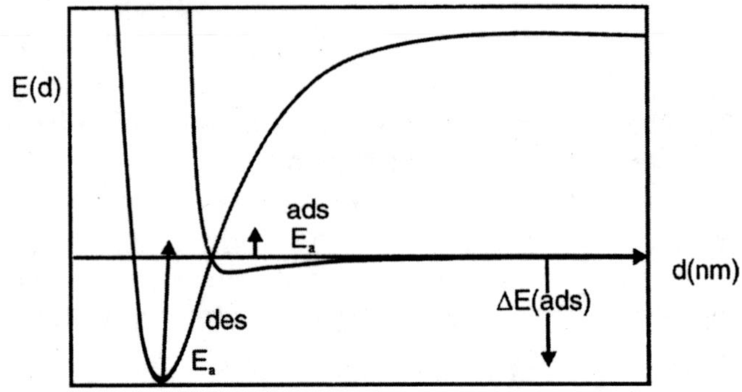


Clearly, from the diagram $E_a^{des} - E_a^{ads} = -DE_{ads}$

but, since the activation energy for adsorption is nearly always very much smaller than that for desorption, and the difference between the energy and enthalpy of adsorption is also very small, it is still quite common to see the relationship

$E_a^{des} \sim -DH_{ads}$

As indicated in the previous section there are many more parameters which affect the energy of the system and the adsorption characteristics in addition to the distance (d) of the centre of mass of the molecule from the surface. We will now look at several examples which illustrate these effects.

## ADSORPTION OF CO ON METAL SURFACES

Some of the additional factors which need to be considered are:

- When carbon monoxide is molecularly adsorbed onto metal surfaces, then it bonds to the surface through the carbon atom. This would imply that the orientation of the molecule as it approaches the surface must be a factor in determining the probability of adsorption and the energy of interaction.

- Spectroscopic studies show that CO adopts a range of bonding geometries in which it is preferentially coordinated to one, two or three metal atoms. This indicates that the lateral position of the molecule must, at the very least, affect the depth of the chemisorption well.
- In many CO/metal systems there are strong coverage dependent effects - these may be associated with changes in the bonding geometry with coverage and/ or intermolecular interactions between the adsorbed CO molecules at high coverages.
- Very reactive surfaces may induce immediate CO dissociation; on other surfaces CO may be predominantly adsorbed in a molecular form but dissociation can be induced by an increase in the substrate temperature, providing the approaching gas phase molecule with high translational or vibrational energy, or at particular surface sites (e.g. step sites).

All these points indicate that a simple 1D PE curve can only provide a very rough first approximation to the interaction of the molecule with the surface. Indeed the situation would be hopelessly complicated if we attempted to take all these effects into account, so instead we will revert to a simpler system and consider just a couple of additional factors.

## ADSORPTION OF HYDROGEN ON W(100)

This system retains some simplifying elements, namely that:

- The chemisorption of hydrogen on tungsten is entirely dissociative and, as noted previously, the physisorption well is extremely shallow.
- The ideal (100) surface plane of tungsten (a bcc metal) has a regular square array of tungsten atoms.

And we will only look at the zero coverage limit (i.e. adsorption on the clean surface) with the molecule approaching along the surface normal. We will, however, now extend the 1D model to consider the effect of two additional variables as the molecule approaches the surface

1. The angular orientation of the molecule (with the limitation that the molecular axis will only be permitted to be either parallel or perpendicular to the surface plane)
2. The lateral position on the surface (with the limitation that the molecule will be considered either to be approaching directly towards a W atom (approaching the "atop site") or directly towards a "bridging site" located between two surface W atoms)

The four slices though the 6D potential energy surface (with energies in eV) where the y-axis in each case is the distance between the surface plane and the molecular centre-of-mass, and the x-axis is the H-H separation.

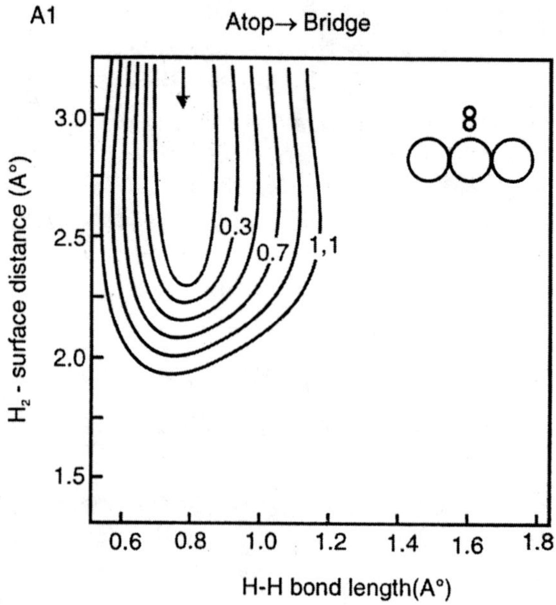


A. Hydrogen molecule approaching the W Surface end-on.

A1. When the hydrogen molecule approaches end-on to an atop site (i.e. from directly above a W) the potential energy increases rapidly for a molecule/surface separation of less than 2.3 Å. The molecule is therefore repelled from the surface.

A2. When the hydrogen molecule approaches end-on to a bridging site the potential energy increases slowly and although some lengthening of the H-H bond may occur with little corresponding increase in energy, there is no facile dissociation channel and the molecule is weakly repelled from the surface.

B. Hydrogen molecule approaching the W Surface broadside-on.

B1. When the hydrogen molecule approaches broadside-on to an atop site then the molecule may dissociate with considerable energy gain and there is no activation barrier to this process (i.e. the potential energy surface is strongly attractive). The deep well in the lower right corner corresponds to the hydrogen atoms produced by dissociation entering the bridge sites of the W surface.

B2. When the hydrogen molecule approaches broadside-on to a bridge site then the molecule may again dissociate but the energy gain as the hydrogen atoms enter the atop sites is less marked than in the previous case (B1) and there is an activation barrier of ca. 0.2 eV that must be overcome for dissociation to occur.

## Adsorbate Geometries and Structures on Metals

We can address the question of what happens when a molecule becomes adsorbed onto a surface at two levels; specifically we can aim to identify

1. The nature of the adsorbed species and its local adsorption geometry (i.e. its chemical structure and co-ordination to adjacent substrate atoms)
2. The overall structure of the extended adsorbate/ substrate interface (i.e. the long range ordering of the surface)

The latter topic is covered in detail in: Classification of Overlayer Structures whist this section will consider only the local adsorption geometry and adsorbate structure.

Chemisorption, by definition, involves the formation of new chemical bonds between the adsorbed species and the surface atoms of the substrate - basically the same type of bonds that are present in any molecular complex. In considering what type of species may be formed on a metal surface, therefore, it is important not to abandon chemical common sense and, if in doubt, to look for inspiration at the structures of known metal-organic complexes.

## CHEMISORPTION OF HYDROGEN AND HALOGENS

### Hydrogen ($H_2$)

In the $H_2$ molecule, the valence electrons are all involved in the H-H s-bond and there are no additional electrons which may interact with the substrate atoms. Consequently, chemisorption of hydrogen on metals is almost invariably a dissociative process in which the H-H bond is broken, thereby permitting the hydrogen atoms to independently interact with the substrate (a description of the energetics of this process). The adsorbed species in this instance therefore are hydrogen atoms.

The exact nature of the adsorbed hydrogen atom complex is generally difficult to determine experimentally, and the very small size of the hydrogen atom does mean that migration of hydrogen from the interface into sub-surface layers of the substrate can occur with relative ease on some metals (Pd, rare earth metals).

The possibility of molecular $H_2$ chemisorption at low temperatures cannot be entirely excluded, however, as demonstrated by the discovery of molecular hydrogen transition metal compounds, such as $W(h^2\text{–}H_2)(CO)_3(Pi\text{–}Pr_3)_3$, in which both atoms of the hydrogen molecule are coordinated to a single metal centre.

### Halogens ($F_2$, $Cl_2$, $Br_2$)

Halogens also chemisorb in a dissociative fashion to give adsorbed halogen atoms. The reasons for this are fairly clear - in principle a halogen molecule could act as a Lewis base and bind to the surface without breakage of the X-X bond, in practice the lone pairs are strongly held by the highly electronegative halogen atom so any such interaction would be very weak and the thermodynamics lie very heavily in favour of dissociative adsorption [i.e. D(X-X) + D(M-$X_{2)}$ << 2 D(M-X)]. Clearly the kinetic barrier to dissociation must also be low or non-existent for the dissociative adsorption to occur readily.

Another way of looking at the interaction of a halogen molecule with a metal surface is as follows: the significant difference in electronegativity between a typical metal and halogen is such that substantial electron transfer from the metal to halogen is favoured. If a halogen molecule is interacting with a metal surface then this transferred electron density will enter the s* antibonding orbital of the molecule, thereby weakening the X-X bond. At the same time the build-up of negative charge on the halogen atoms enhances the strength of the metal-halogen interaction. The net result of these two effects when taken to their limit is that the halogen molecule dissociates and the halogen atoms interact with the metal with a strong ionic contribution to the bonding.

This behaviour is typical of atomic adsorbates which almost invariably endeavour to maximise their co-ordination and hence prefer to occupy the highest-available co-ordination site on the surface.

As a result of the electron transfer from the metal substrate to the halogen atoms, each adsorbed atom is associated with a significant surface dipole.

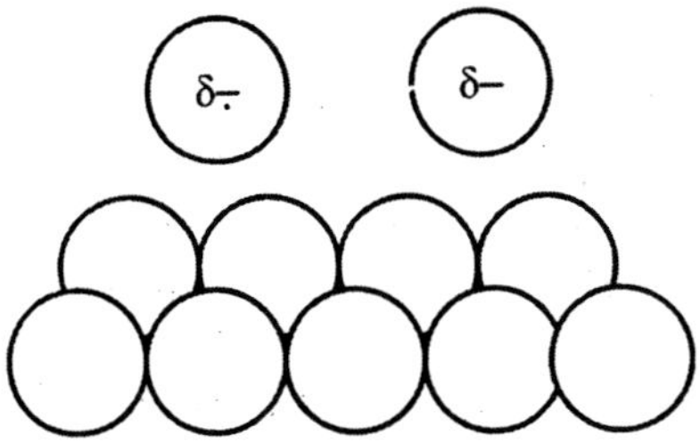


**Fig.** Cross-section

One consequence of this is that there are repulsive (dipole-dipole) interactions between the adsorbed atoms, which are especially evident at higher surface coverages and which can lead to a substantial reduction in the enthalpy of adsorption at specific coverages (if these coverages mark a watershed, above which the atoms are forced to occupy sites which are much closer together).

Another feature of the halogen adsorption chemistry of some metals is the transition from an adsorbed surface layer to surface compound formation at high gas exposures.

## CHEMISORPTION OF NITROGEN AND OXYGEN

### Oxygen

Oxygen is an example of a molecule which usally adsorb dissociatively, but are also found to adsorb molecularly on some metals (e.g. Ag, Pt). In those cases where both types of adsorption are observed it is the dissociative process that corresponds to the higher adsorption enthalpy.

The molecular adsorption state the interaction between the molecule and the surface is relatively weak. Molecules aligned such that the internuclear axis is parallel to the surface plane may bond to a single metal atom of the surface via both

1. $\sigma$-donor interaction, in which the charge transfer is from the occupied molecular $\pi$-bonding molecular orbital of the molecule into vacant orbitals of $\sigma$-symmetry on the metal (i.e. $M \leftarrow O_2$), and

2. p-acceptor interaction, in which an occupied metal *d*-orbital of the correct symmetry overlaps with empty $\pi^*$ orbitals of the molecule and the charge transfer is from the surface to the molecule (i.e. $M \rightarrow O_{2)}$.

Although the interaction of the molecule with the surface is generally weak, one might expect that there might be a substantial barrier to dissociation due to the high strength (and high dissociation enthalpy) of the O=O bond. Nevertheless on most metal surfaces, dissociation of oxygen is observed to be facile which is related to the manner in which the interaction with the surface can mitigate the high intrinsic bond energy and thereby facilitate dissociation.

Once formed, oxygen atoms are strongly bound to the surface and, as noted previously, will tend to occupy the highest available co-ordination site. The strength of the interaction between adsorbate and substrate is such that the adjacent metal atoms are often seen to undergo significant displacements from the equilibrium positions that they occupy on the clean metal surface. This displacement may simply lead to a distortion of the substrate surface in the immediate vicinity of the adsorbed atom or to a more extended surface reconstruction.

Dissociative oxygen adsorption is frequently irreversible - rather than simply leading to desorption, heating of an adsorbed oxygen overlayer often results in either the gradual removal of oxygen from the surface by diffusion into the bulk of the substrate (e.g. Si(111) or Cu(111)) or to the formation of a surface oxide compound. Even at ambient temperatures, extended oxygen exposure often leads to the nucleation of a surface oxide. Depending on the reactivity of the metal concerned, further exposure at low temperatures may result either in a progressive conversion of the bulk material to oxide or the oxidation process may effectively stop after the formation of a passivating surface oxide film of a specific thickness (e.g. Al).

## Nitrogen

The interaction of nitrogen with metal surfaces shows many of the same characteristics as those described above for oxygen. However, in general $N_2$ is less susceptible to dissociation as a result of the lower M-N bond strength and the substantial kinetic barrier associated with breaking the NºN triple bond.

## Chemisorption of Carbon Monoxide

Depending upon the metal surface, carbon monoxide may adsorb either in a molecular form or in a dissociative fashion

- in some cases both states coexist on particular surface planes and over specific ranges of temperature.

1. On the reactive surfaces of metals from the left-hand side of the periodic table (Na, Ca, Ti, rare earth metals) the adsorption is almost invariably dissociative, leading to the formation of adsorbed carbon and oxygen atoms (and thereafter to the formation of surface oxide and oxy-carbide compounds).
2. By contrast, on surfaces of the metals from the right hand side of the *d*-block (Cu, Ag) the interaction is predominantly molecular; the strength of interaction between the CO molecule and the metal is also much weaker, so the M-CO bond may be readily broken and the CO desorbed from the surface by raising the surface temperature without inducing any dissociation of the molecule.
3. For the majority of the transition metals, however, the nature of the adsorption is very sensitive to the surface temperature and surface structure.

Molecularly chemisorbed CO has been found to bond in various ways to single crystal metal surfaces - analogous to its behaviour in isolated metal carbonyl complexes.

Terminal ("Linear") (all surfaces)

Bridging (2f - site) (all surfaces)

Bridging / 3f hollow (*fcc*(111))

Bridging / 4f hollow (rare *fcc*(100) ?)

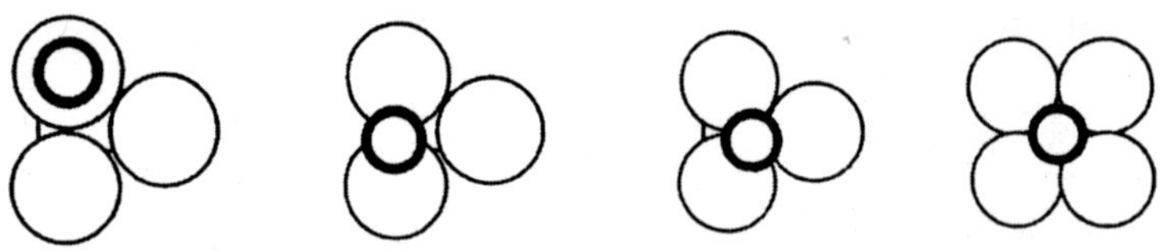

Whilst the above structural diagrams amply demonstrate the inadequacies of a simple valence bond description of the bonding of molecules to surface, they do to an extent also illustrate one of its features and strengths - namely that a given element, in this case carbon, tends to have a specific valence. Consequently, as the number of metal atoms to which the carbon is co-ordinated increases, so there is a corresponding reduction in the C-O bond order.

However, it must be emphasised that a molecule such as CO does not necessarily prefer to bind at the highest available co-ordination site. So, for example, the fact that there are 3-fold hollow sites on an *fcc*(111) surface does not mean that CO will necessarily adopt this site - the preferred site may still be a terminal or 2-fold bridging site, and the site or site(s) which is(are) occupied may change with either surface coverage or temperature. The energy difference between the various adsorption sites available for molecular CO chemisorption appears therefore to be very small.

## CHEMISORPTION OF AMMONIA AND OTHER GROUP V/VI HYDRIDES

Ammonia has lone pairs available for bonding on the central nitrogen atom and may bond without dissociation to a single metal atom of a surface, acting as a Lewis base, to give a pseudo-tetrahedral co-ordination for the nitrogen atom.

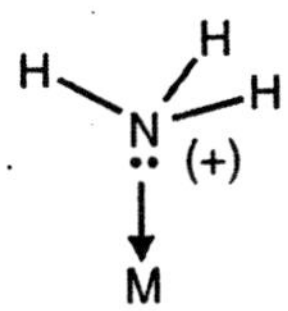


Alternatively, progressive dehydrogenation may occur to give surface $NH_x$ ($x$ = 2,1,0) species and adsorbed hydrogen atoms, i.e.

$$NH_3 \rightarrow NH_{2(ads)} + H_{(ads)} \rightarrow NH_{(ads)} + 2\,H_{(ads)} \rightarrow N_{(ads)} + 3\,H_{(ads)}$$

As the number of hydrogens bonded to the nitrogen atom is reduced, the adsorbed species will tend to move into a higher co-ordination site on the surface (thereby tending to maintain the valence of nitrogen).

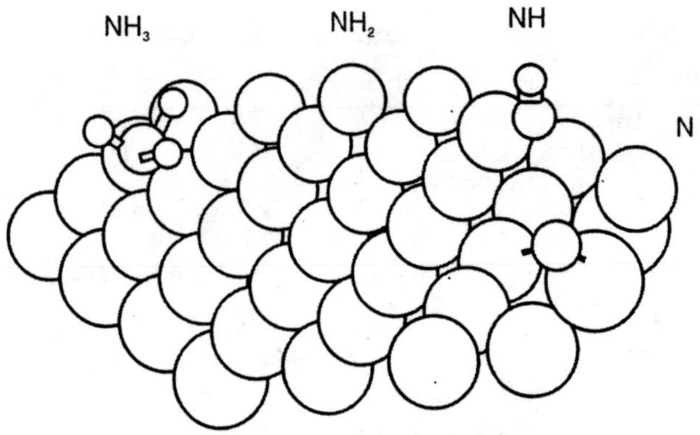


**Fig.** Decomposition fragments of ammonia on an fcc(111) surface

Other Group V and Group VI hydrides ($PH_3$, $H_2O$, $H_2S$) exhibit similar adsorption characteristics to ammonia.

## CHEMISORPTION OF UNSATURATED HYDROCARBONS

Unsaturated hydrocarbons (alkenes, alkynes, aromatic molecules etc.) all tend to interact fairly strongly with metal atom surfaces. At low temperatures (and on less reactive metal surfaces) the adsorption may be molecular, albeit perhaps with some distortion of bond angles around the carbon atom.

Ethene, for example, may bond to give both a p-complex (A) or a di-s adsorption complex (B):

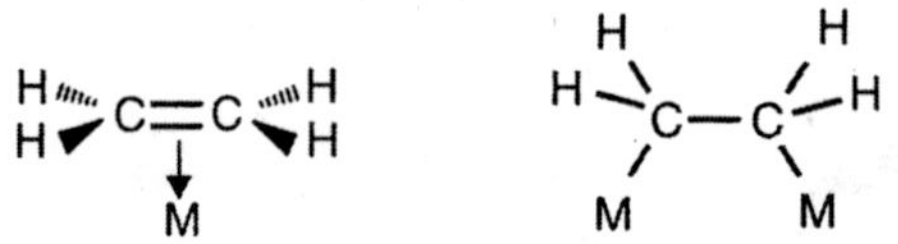

(A) *Chemisorbed* (B) *Ethene*

As the temperature is raised, or even at low temperatures on more reactive surfaces (in particular those that bind hydrogen strongly), a stepwise dehydrogenation may occur. One particularly stable surface intermediate found in the dehydrogenation of ethene is the *ethylidyne* complex, whose formation also involves H-atom transfer between the carbon atoms.

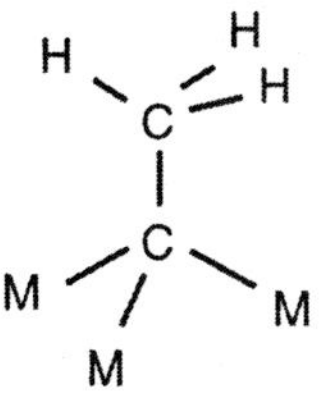


Fig. Ethylidyne: this adsorbate preferentially occupies a 3-fold hollow site to give pseudo-tetrahedral co-ordination for the carbon atom.

The ultimate product of complete dehydrogenation, and the loss of molecular hydrogen by desorption, is usually either carbidic or graphitic surface carbon.

## THE DESORPTION PROCESS

An adsorbed species present on a surface at low temperatures may remain almost indefinitely in that state. As the temperature of the substrate is increased, however, there will come a point at which the thermal energy of the adsorbed species is such that one of several things may occur:

1. A molecular species may decompose to yield either gas phase products or other surface species.
2. An atomic adsorbate may react with the substrate to yield a specific surface compound, or diffuse into the bulk of the underlying solid.
3. The species may desorb from the surface and return into the gas phase.

The last of these options is the desorption process. In the absence of decomposition the desorbing species will generally be the same as that originally adsorbed but this is not necessarily always the case.

(An example where it is not is found in the adsorption of some alkali metals on metallic substrates exhibiting a high work function where, at low coverages, the desorbing species is the alkali metal ion as opposed to the neutral atom. Other examples would include certain isomerisation reactions.)

**Desorption Kinetics**

The rate of desorption, $R_{des,}$ of an adsorbate from a surface can be expressed in the general form:

$$R_{des} = k N^{x}$$

where $x$ - kinetic order of desorption

$k$ - rate constant for the desorption process

$N$ - surface concentration of adsorbed species

The order of desorption can usually be predicted because we are concerned with an elementary step of a "reaction": specifically,

**Atomic or Simple Molecular Desorption**

$$A_{(ads)} \rightarrow A_{(g)}$$

$$M_{(ads)} \rightarrow M_{(g)}$$

- will usually be a first order process (i.e. $x = 1$). Examples include ...

$W / Cu_{(ads)} \rightarrow W_{(s)} + Cu_{(g)}$; desorption of Cu atoms from a W surface

$Cu / CO_{(ads)} \rightarrow Cu_{(s)} + CO_{(g)}$; desorption of CO molecules from a Cu surface

**Recombinative Molecular Desorption**

$$2A_{(ads)} \rightarrow A_{2(g)}$$

will usually be a second order process (i.e. $x = 2$). Examples include ...

$Pt / O_{(ads)} \rightarrow Pt_{(s)} + O_{2\,(g)}$; desorption of O atoms as $O_2$ from a Pt surface

$Ni / H_{(ads)} \rightarrow Ni_{(s)} + H_{2\,(g)}$; desorption of H atoms as $H_2$ from a Ni surface

The rate constant for the desorption process may be expressed in an Arrhenius form,

$$k_{des} = A \exp(-E_a^{des} / RT)$$

where $E_a^{des}$ is the activation energy for desorption, and

*A* is the pre-exponential factor; this can also be considered to be the "attempt frequency", n, at overcoming the barrier to desorption.

This then gives the following general expression for the rate of desorption

$$R_{des} = -dN/dt = v \cdot N^x \cdot \exp\left(-E_a^{des}/RT\right)$$

In the particular case of simple molecular adsorption, the pre-exponential/frequency factor (n) may also be equated with the frequency of vibration of the bond between the molecule and substrate; this is because every time this bond is stretched during the course of a vibrational cycle can be considered an attempt to break the bond and hence an attempt at desorption.

## SURFACE RESIDENCE TIMES

One property of an adsorbed molecule that is intimately related to the desorption kinetics is the *surface residence time* - this is the average time that a molecule will spend on the surface under a given set of conditions (in particular, for a specified surface temperature) before it desorbs into the gas phase.

For a first order process such as the desorption step of a molecularly adsorbed species:

$$M_{(ads)} \rightarrow M_{(g)}$$

the average time (t) prior to the process occurring is given by:

$$t = 1/k_1$$

where $k_1$ is the first order rate constant (no proof given).

From the previously derived desorption formulae we know that

$$k_1 = \text{n} \exp(-E_a^{des}/RT)$$

and if we also substitute for $E_a^{des}$ using the approximate relation $E_a^{des} \sim -DH_{ads}$, then we get the following expression for the surface residence time

$$t = t_o \exp(-DH_{ads}/RT)$$

where: $t_o$ (= 1/n) corresponds to the period of vibration of the bond between the adsorbed molecule and substrate and is frequently taken to be about $10^{-13}$ s.

# 3

# Surface Analytical Techniques

## SURFACE SENSITIVITY AND SURFACE SPECIFICITY

### General Sensitivity Problems

The problems of sensitivity and detection limits are common to all forms of spectroscopy; some techniques are simply better than others in this respect ! In its simplest form, the question of sensitivity boils down to whether it is possible to detect the desired signal above the noise level.

In virtually all surface studies (especially those on single crystal substrates) sensitivity is a major problem. Consider the case of a sample with a surface of size 1 $cm^2$ - this will have ca. $10^{15}$ atoms in the surface layer. In order to detect the presence of impurity atoms present at the 1% level, a technique must be sensitive to ca. $10^{13}$ atoms.

Contrast this with a spectroscopic technique used to analyse a 1 $cm^3$ bulk liquid sample i.e. a sample of ca. $10^{22}$ molecules. The detection of $10^{13}$ molecules in this sample would require 1 ppb (one *part-per-billion*) sensitivity - very few techniques can provide anything like this level of sensitivity !

This is one reason why common spectroscopic techniques such as NMR (detection limit ca. $10^{19}$ molecules) cannot be used for surface studies, except on samples with very high surface areas.

### Surface Sensitivity and Specificity

Assuming that a technique of sufficient sensitivity can be found, another major problem that needs to be addressed in surface spectroscopy is distinguishing between signals from the surface and the bulk of the sample. In principle, there are two ways around this problem:

1. To ensure that the surface signal is distinguishable (shifted) from the comparable bulk signal, and that the detection system has sufficient dynamic range to detect very small signals in the presence of neighbouring large signals.
2. To ensure that the bulk signal is small compared to the surface signal i.e. that the vast majority of detected signal comes from the surface region of the sample.

It is the latter approach which is used by the majority of surface spectroscopic techniques - such techniques can then be said to be *surface sensitive*. To illustrate this, we shall look at one way in which surface sensitivity can be achieved - this makes use of the special properties of

### LOW ENERGY ELECTRONS

it is an approach employed in common surface spectroscopic techniques such as

Auger Electron Spectroscopy (AES)
X-ray Photoelectron Spectroscopy (XPS)

We shall illustrate this approach by considering the latter technique (XPS) but you should remember that the ideas that we develop are more widely applicable.

### XPS - A Surface Sensitive Technique ?

In answering this important question, we will actually address the topics listed below - you are strongly advised to work through them in the order in which they are given.

*Contents:*

- What do we really mean by surface sensitivity / specificity ?

- How can we demonstrate that XPS is surface sensitive ?
- Why is the XPS technique surface sensitive ?
- How can we use knowledge of the IMFP to calculate the thickness of surface films ?

How can we change the degree of surface sensitivity ?

**What do we Really mean by Surface Sensitivity/specificity ?**

Most analytical techniques used in chemistry are "bulk" techniques in the sense that they measure all the atoms within a typical sample (be it a solid, liquid, solution or gas phase sample) - so if, for example, we were to analyse a sample consisting of a 10 nm thick layer of one material (A) deposited on a 1 mm thick substrate of another material (B)....

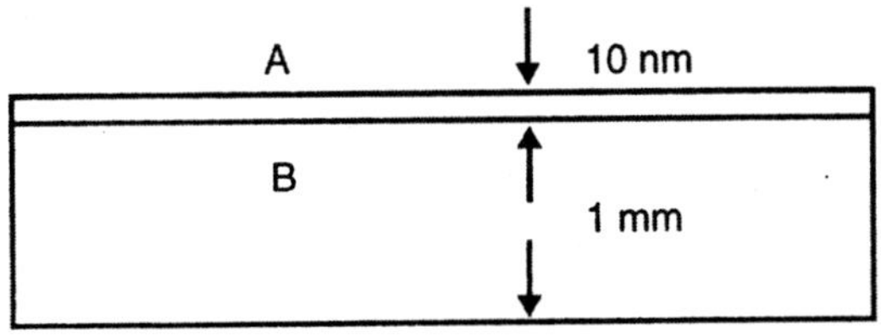


*What would be the concentration of A in parts per million (ppm) which you would expect to be revealed by a bulk analysis of such a sample ?*

.... we would get the same result (10 ppm) as if we were looking at a sample in which component A was uniformly distributed throughout the bulk of B such as to give the same overall concentration in the sample as a whole.

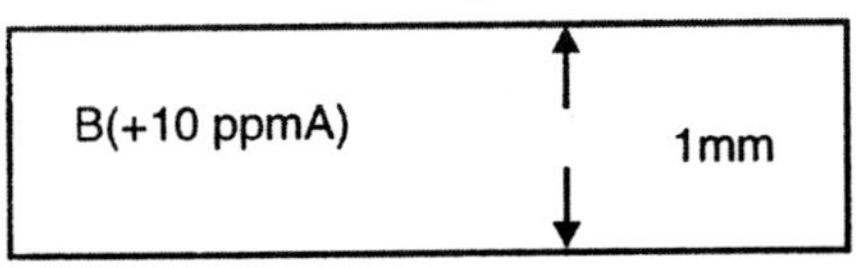


By contrast, a *surface sensitive* technique is more sensitive to those atoms which are located near the surface than it is to atoms in the bulk which are well away from the surface (i.e. the main part of the signal comes from the surface region) - in the case of the first sample therefore a surface sensitive technique would produce a spectrum with enhanced signals due to component A, i.e.

$I_A / I_B >> 10^{-5}$

(where $I_A$ is the signal due to component A etc.)

A truly *surface specific* technique should only give signals due to atoms in the surface region - but that, of course, rather depends how the "surface region" is defined !

Electron spectroscopic techniques such as XPS are not completely surface specific (although you will occasionally find this expression being used) in that whilst most of the signal comes from within a few atomic layers of the surface, a small part of the signal comes from much deeper into the solid - they are best described as being surface sensitive techniques.

However, we should also note that there are a few techniques (e.g. low energy ion scattering [ISS/LEIS]) which are indeed virtually specific to the topmost layer of atoms.

**How can we Demonstrate that XPS is Surface Sensitive ?**

Consider again the first of the two samples considered in the previous section - such samples are often readily prepared by depositing material A by evapouration onto a substrate of material B.

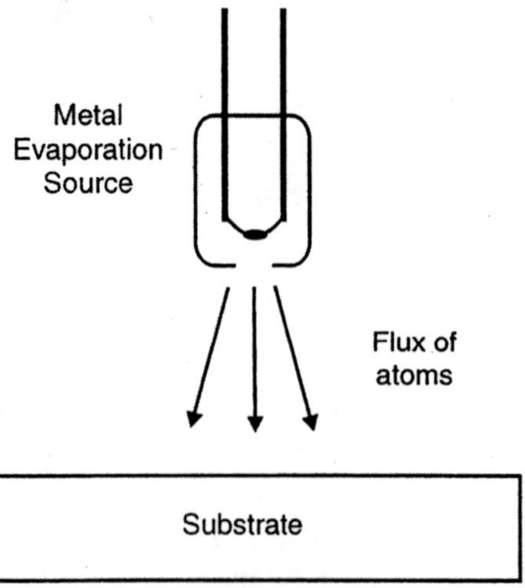


It is experimentally observed that in such a situation the XPS signals due to the substrate (material B) are rapidly attenuated, whilst those due to the condensed evapourant (material A) simultaneously increase to a limiting value.

**Why is the XPS Technique Surface Sensitive ?**

The soft X-rays employed in XPS penetrate a substantial distance into the sample (~mm). Thus this method of excitation imparts no surface sensitivity at the required atomic scale. The surface sensitivity must therefore arise from the emission and detection of the photoemitted electrons.

Consider, therefore, a hypothetical experiment in which electrons of a given energy, $E_{0,}$ are emitted from atoms in a solid at various depths, *d*, below a flat surface.

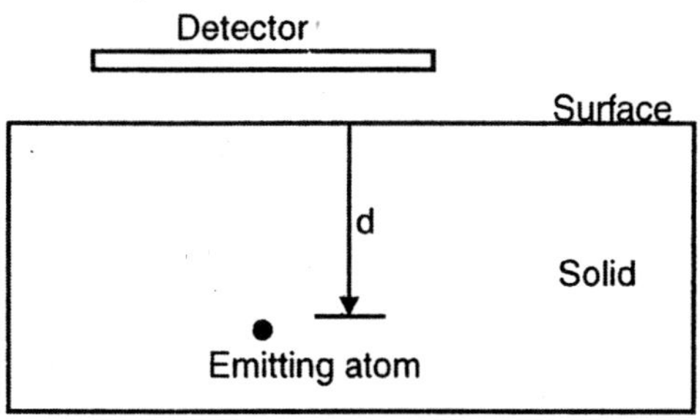


We shall assume that only those electrons which reach the surface and, at the time of leaving the solid, still have their initial energy ($E_{o)}$ are detected in this experiment.

This is not literally true in an XPS experiment, but in XPS experiments it is only those photoelectrons possessing characteristic emission energies and contributing to the peaks and that we consider in the subsequent analysis; we do not consider photoelectrons which have lost some energy and contribute to the background of the spectrum rather than a specific peak.

There are two things which would therefore prevent an emitted electron from being detected:

1. If it were "captured" before reaching the surface, or emitted in the wrong direction and never reached the surface !

2. If it lost energy before reaching the surface, i.e. if it escaped from the solid and reached the detector, but with $E < E_o$

The process by which an electron can lose energy as it travels through the solid is known as "inelastic scattering". Each inelastic scattering event leads to:

1. A reduction in the electron energy
2. A change in the direction of travel.

In the diagram below arrows are used to represent possible trajectories of emitted electrons from a particular source atom. The orange arrows represent electrons travelling with their initial energy, $E_o$, whilst the blue arrows represent electron trajectories subsequent to an inelastic scattering event. Only the orange traces which extend out of the solid therefore correspond to electrons which would be detected in our hypothetical experiment.

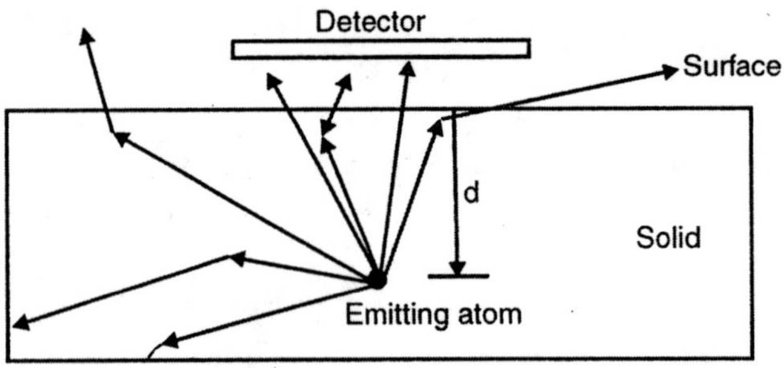


In this case, therefore, only a small fraction of the emitted electrons would have been detected.

Clearly if the source atom had been closer to the surface, then a greater fraction of the emitted electrons would have been detected since:

1. The detector would subtend a greater (solid) angle at the emitting atom (this is actually not a significant factor in real experiments).
2. For an electron emitted towards the surface, there would be less chance of inelastic scattering before it escaped from the solid.

**So how does the Probability of Detection depend upon the distance of the Emitting Atom below the Surface ?**

Let us first simplify the problem by only considering those electrons emitted directly towards (normal to) the surface - in our hypothetical experiment we can represent this by a smaller detector located directly above the emitting atoms.

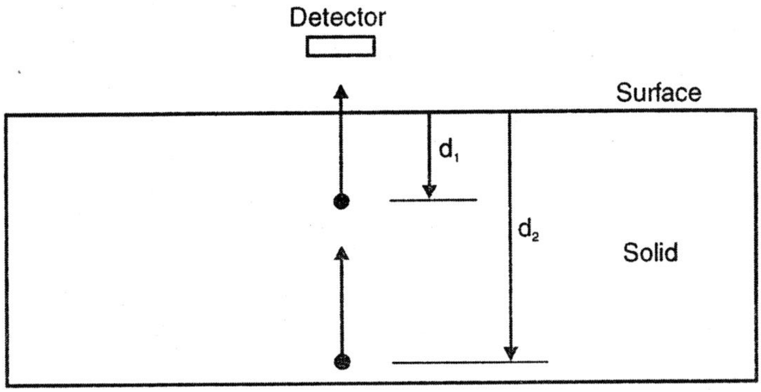


The probability of escape from a given depth, P($d$), is determined by the likelihood of the electron not being inelastically scattered during its journey to the surface.

Clearly, $P(d_2) < P(d_1)$ but, in order to quantify this we need to know more about the inelastic scattering process - this brings us to the subject of the..

Inelastic Mean Free Path (IMFP) of electrons.

The IMFP is a measure of the average distance travelled by an electron through a solid before it is inelastically scattered; it is dependent upon

- The initial kinetic energy of the electron.
- The nature of the solid (but most elements show very similar IMFP v's energy relationships).

The IMFP is actually defined by the following equation which gives the probability of the electron travelling a distance, $d$, through the solid without undergoing scattering

$$P(d) = \exp(-d / \lambda)$$

where l is the IMFP for the electrons of energy $E$.

The graph below illustrates this functional form of P($d$)

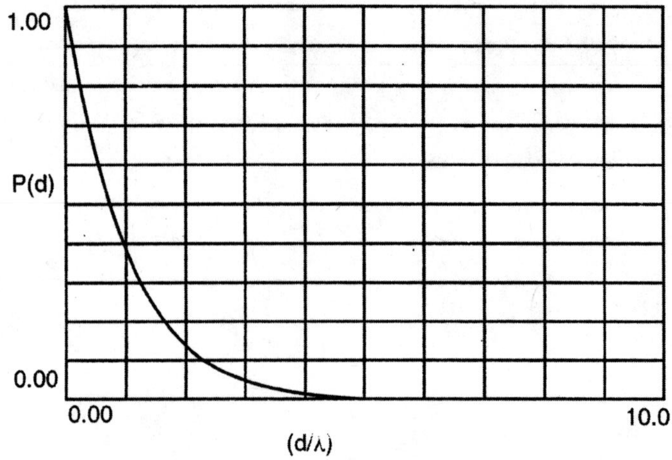


and it is clear that the probability of escape decays very rapidly and is essentially zero for a distance $d$ > 5l. It is therefore useful to re-plot the function for a range of ($d$/l) from 0-5.

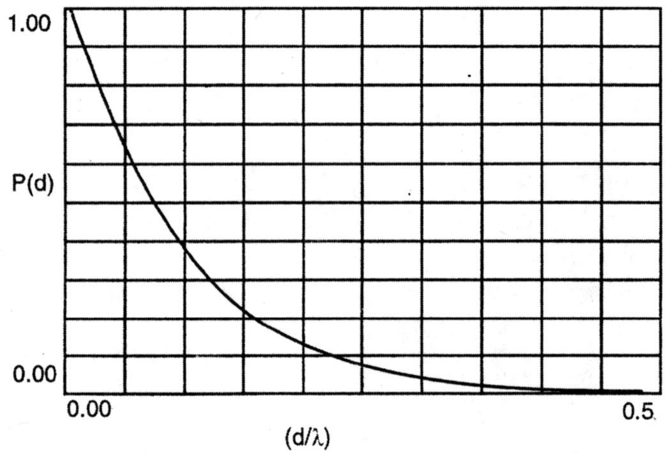


It is also interesting to look at this same problem from a different point of view. Let us assume that there are many sources of electrons uniformly distributed at all distances from the surface of the solid (which is in effect the situation pertaining in all XPS experiments) and again arrange to only detect those unscattered electrons which emerge normal to the surface - then we can ask...

**What is the Distribution of the Depths of Origin of those Electrons that we do Detect ?**

This new function, let's call it $P'(d)$, actually has the same exponential form as $P(d)$ - i.e. the frequency of detection of electrons from different depths in the solid is directly proportional to the probability of electron escape from each depth.

**THE MAJORITY OF ELECTRONS DETECTED COME FROM WITHIN ONE "IMFP" OF THE SURFACE**

Virtually all (> 95%) of the electrons detected come from within 3 IMFPs of the surface.

**So how big is the IMFP of typical electrons within a solid such as a metal ?**

The graph below shows the so-called "universal curve" for the variation of the IMFP with initial electron energy (in eV).

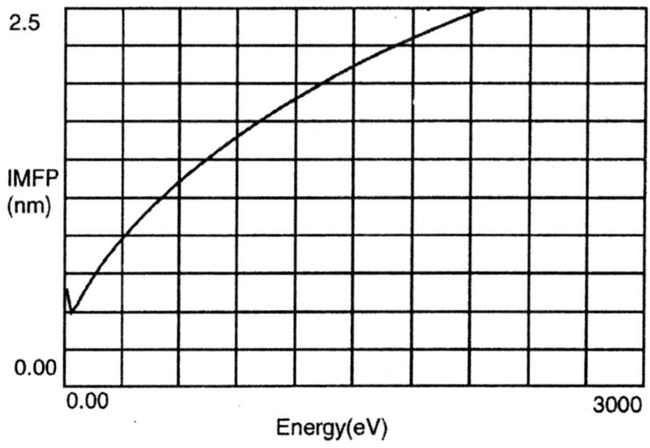


It is reasonably accurate for most metals, although experimental measurements on different metals do show a substantial scatter about the general curve.

Other classes of solids also exhibit qualitatively similar universal curves, although the absolute values of the IMFP of electrons in materials such as inorganic solids and polymers do differ quite significantly from those in metals.

This IMFP data is more usually displayed on a *log-log* plot, since this more clearly shows how the IMFP varies at low kinetic energies - the graph below shows the general curve for metals re-plotted in this new format.

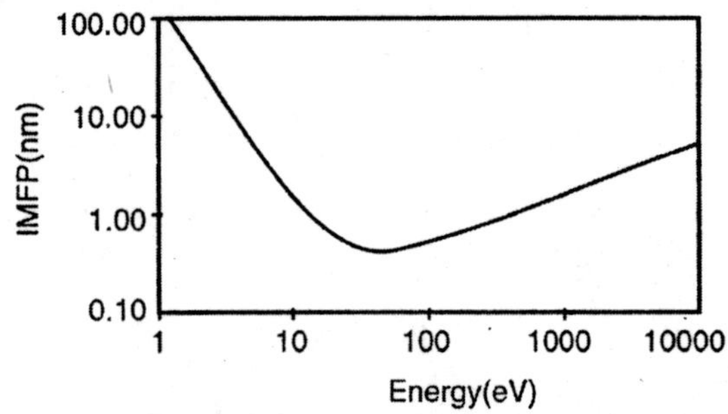


The IMFP exhibits a minimum for electrons with a kinetic energy of around 50 -100 eV; at lower energies the probability of inelastic scattering decreases since the electron has insufficient energy to cause plasmon excitation (the main scattering mechanism in metals), and consequently the distance between inelastic collisions and the IMFP increases.

The Inelastic Mean Free Path (IMFP) in metals is typically less than:

10 Angstroms (1 nm) for electron energies in the range 15 < *E*/eV < 350

20 Angstroms (2 nm) for electron energies in the range 10 < *E*/eV < 1400

i.e. the IMFP of low energy electrons corresponds to only a few atomic layers.

This means that any experimental technique such as XPS which involves the generation and detection of electrons of such energies will be surface sensitive.

**How can we use knowledge of the IMFP to calculate the thickness of surface films ?**

We can now consider a situation where a substrate (or thick film) of one material, B, is covered by a thin film of a different material, A. The XPS signal from the underlying substrate will be attenuated (i.e. reduced in intensity) due to inelastic scattering of some of the photoelectrons as they traverse through the layer of material A.

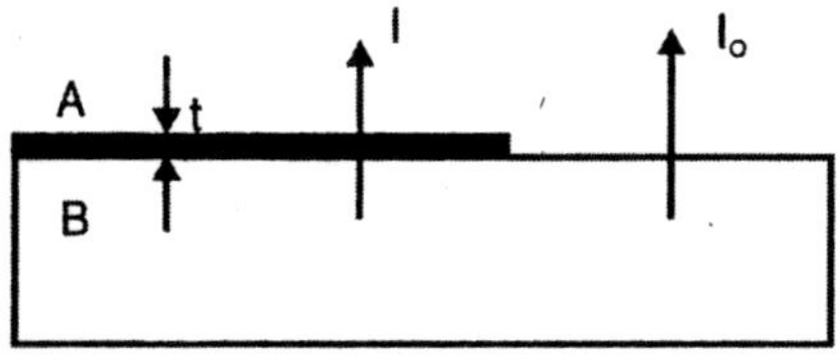


The probability of such a scattering event for any single photoelectron passing through this layer is simply given by:

$$P = \exp(-t / \lambda)$$

where: *t* is the thickness of the layer of material A.

It follows that the overall intensity of a XPS signal arising from B is reduced by this same factor, i.e. if the intensity of this signal in the absence of any covering layer is $I_0$ then the intensity $I$ in the presence of the overlayer is given by:

$$I = I_0 \exp(-t / \lambda)$$

It also follows that it is possible to estimate the thickness of a deposited layer (using the above equation), provided the reduction in the substrate signal is known (i.e. if spectra are acquired before and after deposition of the covering film).

**How can we change the Degree of Surface Sensitivity ?**

For a given technique such as XPS which relies on inelastic electron scattering to provide the surface sensitivity, we can consider how it might be possible to increase the surface sensitivity when analysing a particular substrate.

One approach would be to try and ensure that the electrons being detected have kinetic energies corresponding to the minimum in the universal curve for the IMFP. The problem with this approach is that

1. It requires us to be able to vary the X-ray radiation so that we can adjust the photon energy, *h*n, to give us the desired kinetic energy for the photoemitted electrons; this is not generally possible with conventional laboratory x-ray sources (although tunable x-ray radiation is available at large synchrotron sources)
2. Whilst it may be possible to adjust the kinetic energy of some particular photoelectrons to correspond to the minimum in the universal curve, this effect may not hold for all the other photoelectron peaks associated with the sample.

A much more generally useful approach is to collect photoelectrons at a more grazing emission angle.

The key point to recognise here is that it is actually the distance that the electrons have to travel through the solid (rather than the distance of the emitting atom below the surface) that is crucial in determining their probability of escape without inelastic scattering - only in the special case considered previously of normal emission are these two quantities the same.

Consider, for example, emission at some angle q to the surface normal.

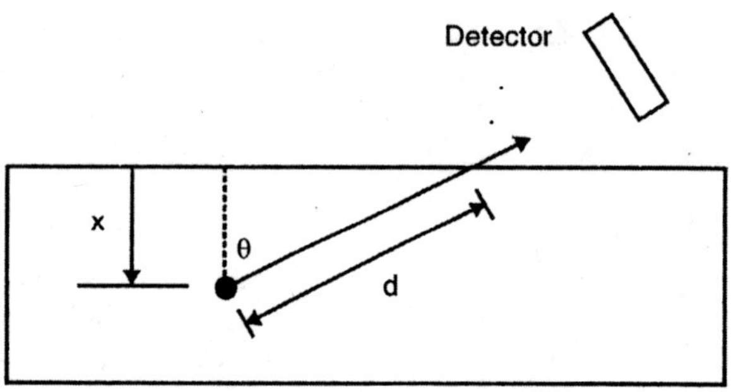


Once again 63% of the electrons will originate from atoms located such that the electrons have to travel a distance, *d*, which is less than 1*l* through the solid. Under these measurement conditions, however, all these atoms are located within a smaller distance, *x*, from the surface where

$x = d \cos \theta$

So as the emission angle ($\theta$) is increased, $\cos \theta$ decreases, the analysed region becomes more surface localised and the surface sensitivity is increased.

One example of how this effect can be put to good use is in the study (and diagnosis) of *surface segregation*. The graph below shows simulated data for the variation of the XPS signal intensities with emission angle for

- A random alloy of 10% of element A in element B, with no segregation (lower blue curve)
- The same random alloy, but with surface segregation of A to give a surface monolayer of pure element A (upper red curve)

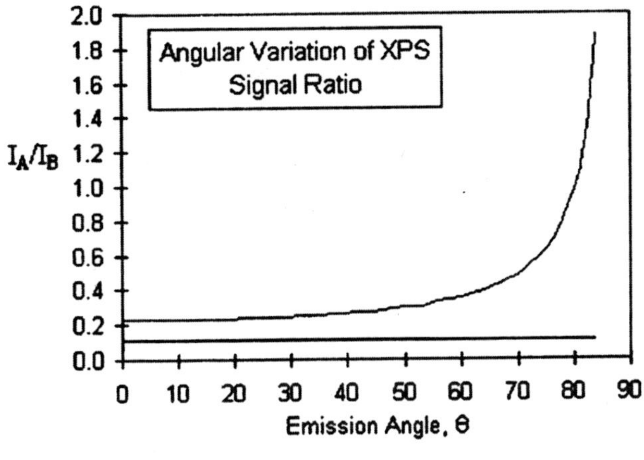


## AUGER ELECTRON SPECTROSCOPY

Auger Electron Spectroscopy (*Auger spectroscopy* or AES) was developed in the late 1960's, deriving its name from the effect first observed by Pierre Auger, a French Physicist, in the mid-1920's. It is a surface specific technique utilising the emission of low energy electrons in the *Auger process* and is one of the most commonly employed surface analytical techniques for determining the composition of the surface layers of a sample.

## PHYSICAL BASIS

Auger spectroscopy can be considered as involving three basic steps:

1. Atomic ionization (by removal of a core electron)
2. Electron emission (the Auger process)
3. Analysis of the emitted Auger electrons

The last stage is simply a technical problem of detecting charged particles with high sensitivity, with the additional requirement that the kinetic energies of the emitted electrons must be determined.

We will therefore concern ourselves only with the first two processes - before starting, however, it is useful to briefly review the electronic structure of atoms and solids, and associated nomenclature.

### Electronic Structure - Isolated Atoms

The diagram below schematically illustrates the energies of the various electron energy levels in an isolated, multi-electron atom, with the conventional chemical nomenclature for these orbitals given on the right hand side.

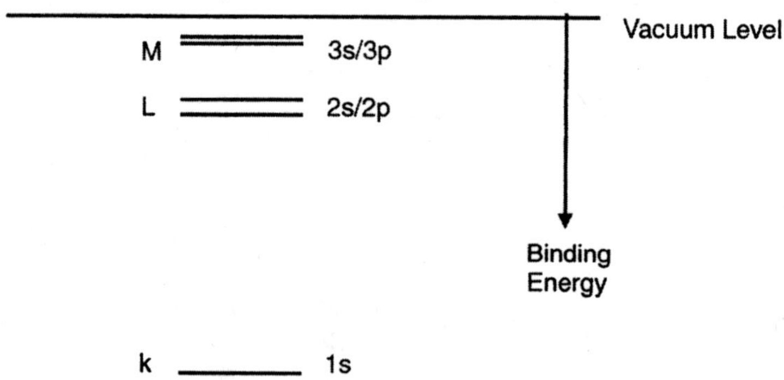


However, scientists working with x-rays tend to use the alternative nomenclature on the left and it is this nomenclature that is used in Auger spectroscopy. The designation of levels to the K,L,M,... shells is based on their having principal quantum numbers of 1,2,3,... respectively.

It is convenient to expand the part of the energy scale close to the vacuum level in order to more clearly distinguish between the higher levels....

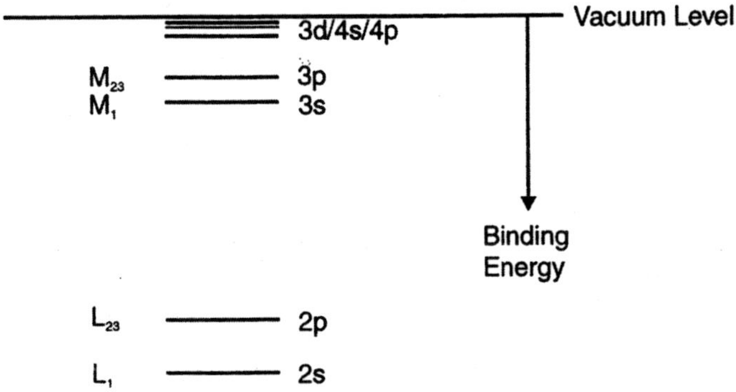


The numerical component of the KLM.. style of nomenclature is usually written as a subscript immediately following the main shell designation.

Levels with a non-zero value of the orbital angular momentum quantum number ($l > 0$), i.e. p,d,f,.. levels, show spin-orbit splitting. The magnitude of this splitting, however, is too small to be evident on this diagram - hence, the double subscript for these levels (i.e. $L_{2,3}$ represents both the $L_2$ and $L_3$ levels).

**Electronic Structure - Solid State**

In the solid state the core levels of atoms are little perturbed and essentially remain as discrete, localised (i.e. atomic-like) levels. The valence orbitals, however, overlap significantly with those of neighbouring atoms generating bands of spatially-delocalised energy levels.

The energy level diagram for the solid is therefore closely resemblant of that of the corresponding isolated atom, except for the levels closest to the vacuum level.

The diagram below shows the electronic structure of Na metal:

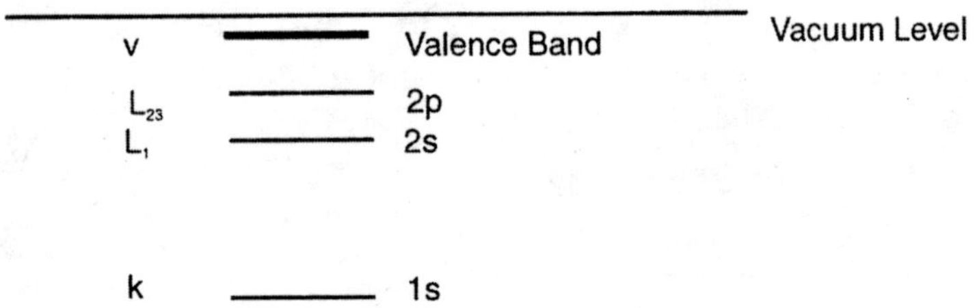


## THE AUGER PROCESS AND AUGER SPECTROSCOPY

Now let us return to the subject of Auger spectroscopy, the Auger process is illustrated using the K, $L_1$ & $L_{2,3}$ levels. These could be the inner core levels of an atom in either a molecular or solid-state environment.

### Ionization

The Auger process is initiated by creation of a core hole - this is typically carried out by exposing the sample to a beam of high energy electrons (typically having a primary energy in the range 2 – 10 keV). Such electrons have sufficient energy to ionise all levels of the lighter elements, and higher core levels of the heavier elements.

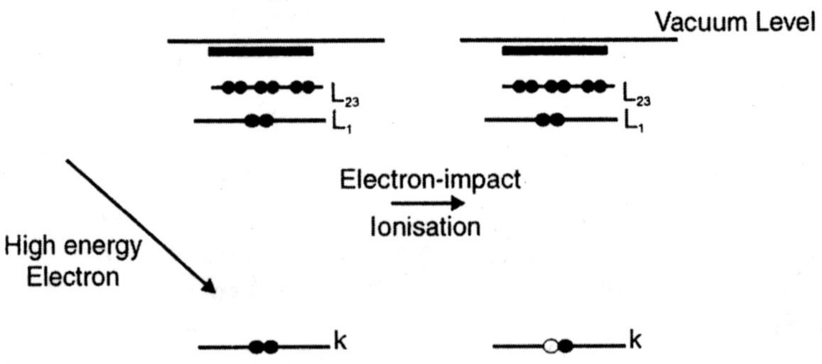


In the diagram above, ionisation is shown to occur by removal of a K-shell electron, but in practice such a crude method of ionisation will lead to ions with holes in a variety of inner shell levels.

In some studies, the initial ionisation process is instead carried out using soft x-rays (hn = 1000 - 2000 eV). In this case, the acronym XAES is sometimes used. As we shall see, however, this change in the method of ionisation has no significant effect on the final Auger spectrum.

**Relaxation and Auger Emission**

The ionized atom that remains after the removal of the core hole electron is, of course, in a highly excited state and will rapidly relax back to a lower energy state by one of two routes:

1. X-ray fluorescence, or
2. Auger emission

We will only consider the latter mechanism, an example of which is illustrated schematically below....

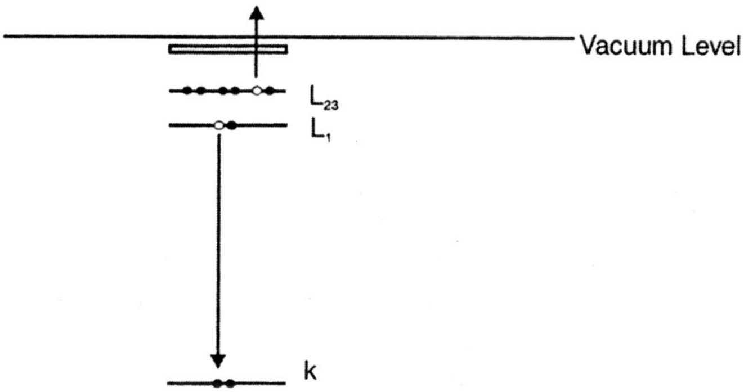


In this example, one electron falls from a higher level to fill an initial core hole in the K-shell and the energy liberated in this process is simultaneously transferred to a second electron; a fraction of this energy is required to overcome the binding energy of this second electron, the remainder is retained by this emitted *Auger electron* as kinetic energy. In the Auger process illustrated, the final state is a doubly-ionized atom with core holes in the $L_1$ and $L_{2,3}$ shells.

We can make a rough estimate of the KE of the Auger electron from the binding energies of the various levels involved. In this particular example,

$$KE = (E_K - E_{L1}) - E_{L23}$$

*the KE of the Auger electron is independent of the mechanism of initial core hole formation.*

The expression for the energy can also be re-written in the form:

$KE = E_K - (E_{L1} + E_{L23})$

It should be clear from this expression that the latter two energy terms could be interchanged without any effect - i.e. it is actually impossible to say which electron fills the initial core hole and which is ejected as an Auger electron; they are indistinguishable.

**An Auger transition is therefore characterized primarily by:-**

1. The location of the initial hole
2. The location of the final two holes

although the existence of different electronic states (terms) of the final doubly-ionized atom may lead to fine structure in high resolution spectra.

When describing the transition, the initial hole location is given first, followed by the locations of the final two holes in order of decreasing binding energy.

i.e. the transition illustrated is a $KL_1L_{2,3}$ transition.

If we just consider these three electronic levels there are clearly several possible Auger transitions: specifically,

$K\,L_1\,L_1 K\,L_1\,L_{2,3}$ $K\,L_{2,3}\,L_{2,3}$

In general, since the initial ionisation is non-selective and the initial hole may therefore be in various shells, there will be many possible Auger transitions for a given element - some weak, some strong in intensity.

Auger Spectroscopy is based upon the measurement of the kinetic energies of the emitted electrons. Each element in a sample being studied will give rise to a characteristic spectrum of peaks at various kinetic energies.

This is an Auger spectrum of Pd metal - generated using a 2.5 keV electron beam to produce the initial core vacancies and hence to stimulate the Auger emission process. The main peaks for palladium occur between 220 & 340 eV. The peaks are situated on a high background which arises from the vast number of so-called *secondary electrons* generated by a multitude of inelastic scattering processes.

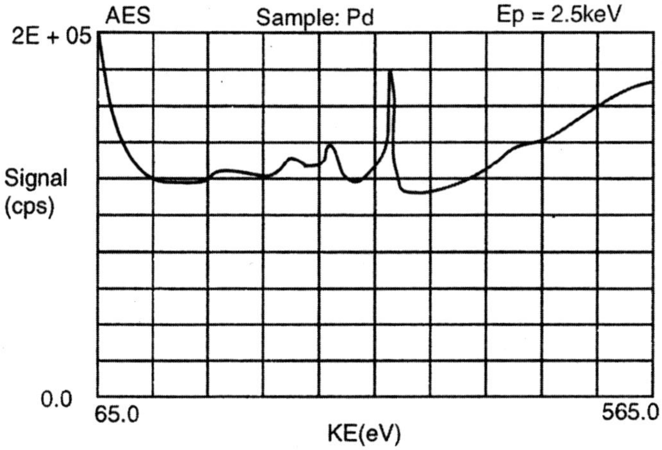


Auger spectra are also often shown in a differentiated form: the reasons for this are partly historical, partly because it is possible to actually measure spectra directly in this form and by doing so get a better sensitivity for detection. The plot below shows the same spectrum in such a differentiated form.

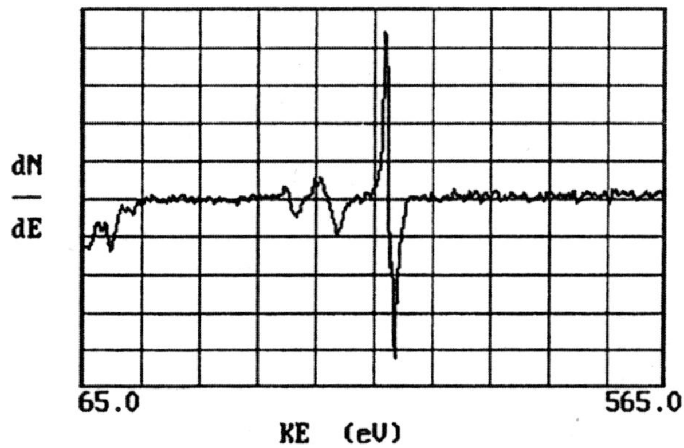


Auger Electron Spectroscopy (AES) is a surface-sensitive spectroscopic technique used for elemental analysis of surfaces; it offers

- High sensitivity (typically ca. 1% monolayer) for all elements except H and He.
- A means of monitoring surface cleanliness of samples
- Quantitative compositional analysis of the surface region of specimens, by comparison with standard samples of known composition.

In addition, the basic technique has also been adapted for use in:

1. *Auger Depth Profiling*: providing quantitative compositional information as a function of depth below the surface.
2. *Scanning Auger Microscopy (SAM)*: providing spatially-resolved compositional information on heterogeneous samples.

## PHOTOELECTRON SPECTROSCOPY

Photoelectron spectroscopy utilizes photo-ionization and energy-dispersive analysis of the emitted photoelectrons to study the composition and electronic state of the surface region of a sample.

Traditionally, when the technique has been used for surface studies it has been subdivided according to the source of exciting radiation into:

| | |
|---|---|
| X-ray Photoelectron Spectroscopy (XPS) | using soft x-ray (200-2000 eV) radiation to examine core-levels. |
| Ultraviolet Photoelect ron Spectroscopy(UPS) | using vacuum UV (10-45 eV) radiation to examine valence levels. |

The development of synchrotron radiation sources has enabled high resolution studies to be carried out with radiation spanning a much wider and more complete energy range (5 - 5000+ eV) but such work is, and will remain, a very small minority of all photoelectron studies due to the expense, complexity and limited availability of such sources.

**Physical Basis**

Photoelectron spectroscopy is based upon a single photon in/electron out process and from many viewpoints this underlying process is a much simpler phenomenon than the Auger process.

The energy of a photon is given by the Einstein relation:

$$E = h\text{n}$$

where $h$ - Planck constant (6.62 x $10^{-34}$ J s)

n - frequency (Hz) of the radiation

Photoelectron spectroscopy uses monochromatic sources of radiation (i.e. photons of fixed energy).

In XPS the photon is absorbed by an atom in a molecule or solid, leading to ionization and the emission of a core (inner shell) electron. By contrast, in UPS the photon interacts with valence levels of the molecule or solid, leading to ionisation by removal of one of these valence electrons.

The kinetic energy distribution of the emitted photoelectrons (i.e. the number of emitted photoelectrons as a function of their kinetic energy) can be measured using any appropriate electron energy analyser and a photoelectron spectrum can thus be recorded.

The process of photoionization can be considered in several ways: one way is to look at the overall process as follows:

$$\text{A} + h\text{n} \rightarrow \text{A}^+ + \text{e}^-$$

Conservation of energy then requires that:

$$\text{E(A)} + h\text{n} = \text{E(A}^{+)}) + \text{E(e}^-)$$

Since the electron's energy is present solely as kinetic energy (KE) this can be rearranged to give the following expression for the KE of the photoelectron:

$$\text{KE} = h\text{n} - (\text{E(A}^{+)}) - \text{E(A))}$$

The final term in brackets, representing the difference in energy between the ionized and neutral atoms, is generally called the *binding energy* (BE) of the electron - this then leads to the following commonly quoted equation:

$$KE = hn - BE$$

An alternative approach is to consider a one-electron model along the lines of the following pictorial representation; this model of the process has the benefit of simplicity but it can be rather misleading.

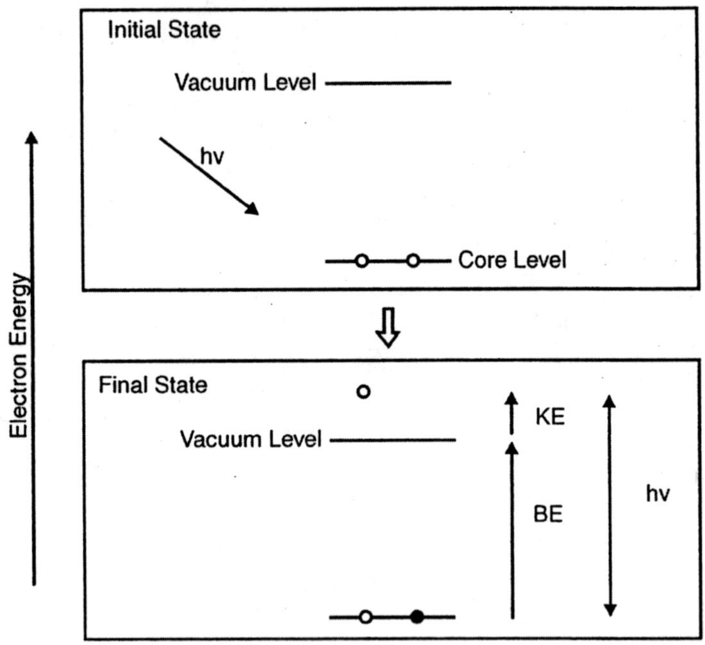


The BE is now taken to be a direct measure of the energy required to just remove the electron concerned from its initial level to the vacuum level and the KE of the photoelectron is again given by:

$$KE = hn - BE$$

The binding energies (BE) of energy levels in solids are conventionally measured with respect to the Fermi-level of the solid, rather than the vacuum level. This involves a small correction to the equation given above in order to account for the *work function* (Φ) of the solid.

**Experimental Details**

The basic requirements for a photoemission experiment (XPS or UPS) are:

1. A source of fixed-energy radiation (an x-ray source for XPS or, typically, a He discharge lamp for UPS)
2. An electron energy analyser (which can disperse the emitted electrons according to their kinetic energy, and thereby measure the flux of emitted electrons of a particular energy)
3. A high vacuum environment (to enable the emitted photoelectrons to be analysed without interference from gas phase collisions)

Such a system is illustrated schematically below:

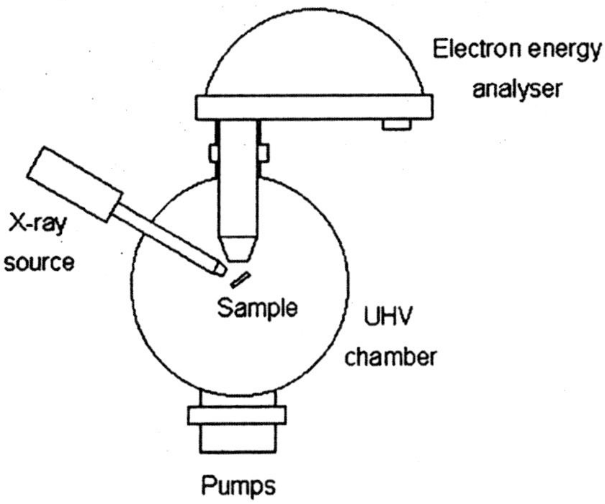


There are many different designs of electron energy analyser but the preferred option for photoemission experiments is a concentric hemispherical analyser (CHA) which uses an electric field between two hemispherical surfaces to disperse the electrons according to their kinetic energy.

**X-RAY PHOTOELECTRON SPECTROSCOPY (XPS)**

For each and every element, there will be a characteristic binding energy associated with each core atomic orbital i.e.

each element will give rise to a characteristic set of peaks in the photoelectron spectrum at kinetic energies determined by the photon energy and the respective binding energies.

The presence of peaks at particular energies therefore indicates the presence of a specific element in the sample under study - furthermore, the intensity of the peaks is related to the concentration of the element within the sampled region. Thus, the technique provides a *quantitative analysis of the surface composition* and is sometimes known by the alternative acronym, ESCA (Electron Spectroscopy for Chemical Analysis).

The most commonly employed x-ray sources are those giving rise to:

Mg $K_{\alpha}$ radiation: $hn = 1253.6$ eV

Al $K_{\alpha}$ radiation: $hn = 1486.6$ eV

The emitted photoelectrons will therefore have kinetic energies in the range of ca. 0 - 1250 eV or 0 - 1480 eV. Since such electrons have very short IMFPs in solids, the technique is necessarily surface sensitive.

**The XPS spectrum of Pd Metal**

The diagram below shows a real XPS spectrum obtained from a Pd metal sample using Mg $K_a$ radiation

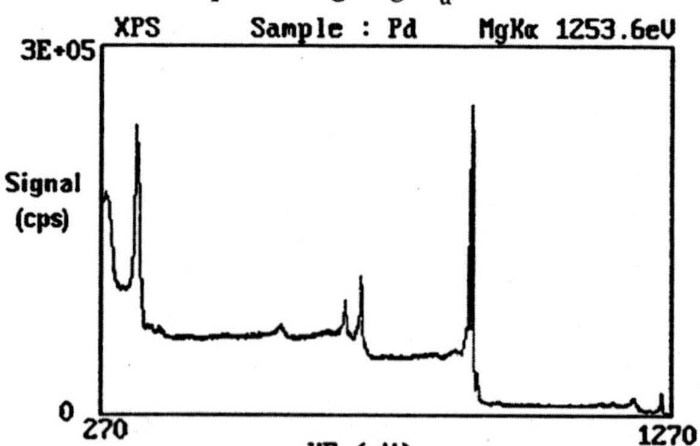


The main peaks occur at kinetic energies of ca. 330, 690, 720, 910 and 920 eV.

Since the energy of the radiation is known it is a trivial matter to transform the spectrum so that it is plotted against BE as opposed to KE.

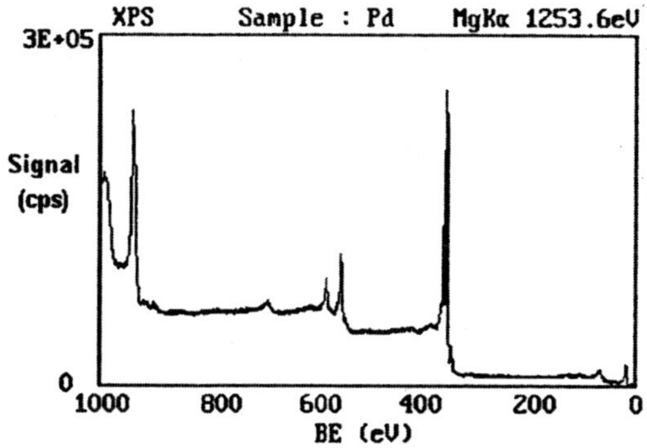


The most intense peak is now seen to occur at a binding energy of ca. 335 eV

Working downwards from the highest energy levels.

1. The valence band (4*d*,5*s*) emission occurs at a binding energy of ca. 0 – 8 eV (measured with respect to the Fermi level, or alternatively at ca. 4 – 12 eV if measured with respect to the vacuum level).
2. The emission from the 4*p* and 4*s* levels gives rise to very weak peaks at 54 and 88 eV respectively
3. The most intense peak at ca. 335 eV is due to emission from the 3*d* levels of the Pd atoms, whilst the 3*p* and 3*s* levels give rise to the peaks at ca. 534/561 eV and 673 eV respectively.
4. The remaining peak is not an XPS peak at all ! - it is an Auger peak arising from x-ray induced Auger emission. It occurs at a kinetic energy of ca. 330 eV (in this case it is really meaningless to refer to an associated binding energy).

It may be further noted that

- There are significant differences in the natural widths of the various photoemission peaks
- The peak intensities are not simply related to the electron occupancy of the orbitals

## Spin-Orbit Splitting

Closer inspection of the spectrum shows that emission from some levels (most obviously $3p$ and $3d$) does not give rise to a single photoemission peak, but a closely spaced doublet.

We can see this more clearly if, for example, we expand the spectrum in the region of the $3d$ emission...

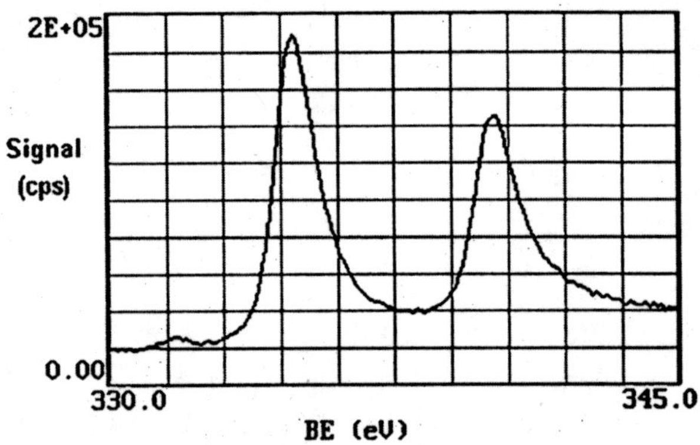


The $3d$ photoemission is in fact split between two peaks, one at 334.9 eV BE and the other at 340.2 eV BE, with an intensity ratio of 3:2. This arises from spin-orbit coupling effects in the final state. The inner core electronic configuration of the initial state of the Pd is:

$(1s)^2\ (2s)^2\ (2p)^6\ (3s)^2\ (3p)^6\ (3d)^{10}$....

with all sub-shells completely full.

The removal of an electron from the $3d$ sub-shell by photo-ionization leads to a $(3d)^9$ configuration for the final state - since the $d$-orbitals ($l = 2$) have non-zero orbital angular momentum, there will be coupling between the unpaired spin and orbital angular momenta.

Spin-orbit coupling is generally treated using one of two models which correspond to the two limiting ways in which the coupling can occur - these being the *LS* (or Russell-Saunders) coupling approximation and the *j-j* coupling approximation.

If we consider the final ionised state of Pd within the Russell-Saunders coupling approximation, the $(3d)^9$ configuration gives rise to two states (ignoring any coupling with valence levels) which differ slightly in energy and in their degeneracy...

$$^2D_{5/2}\ g_J = 2 \times \{5/2\}+1 = 6$$

$$^2D_{3/2}\ g_J = 2 \times \{3/2\}+1 = 4$$

These two states arise from the coupling of the $L = 2$ and $S = 1/2$ vectors to give permitted $J$ values of 3/2 and 5/2. The lowest energy final state is the one with maximum $J$ (since the shell is more than half-full), i.e. $J = 5/2$, hence this gives rise to the "lower binding energy" peak. The relative intensities of the two peaks reflects the degeneracies of the final states ($g_J = 2J + 1$), which in turn determines the probability of transition to such a state during photoionization.

The Russell-Saunders coupling approximation is best applied only to light atoms and this splitting can alternatively be described using individual electron *l-s* coupling. In this case the resultant angular momenta arise from the single hole in the *d*-shell; a *d*-shell electron (or hole) has $l = 2$ and $s = 1/2$, which again gives permitted *j*-values of 3/2 and 5/2 with the latter being lower in energy.

The peaks themselves are conventionally annotated as indicated - note the use of lower case lettering

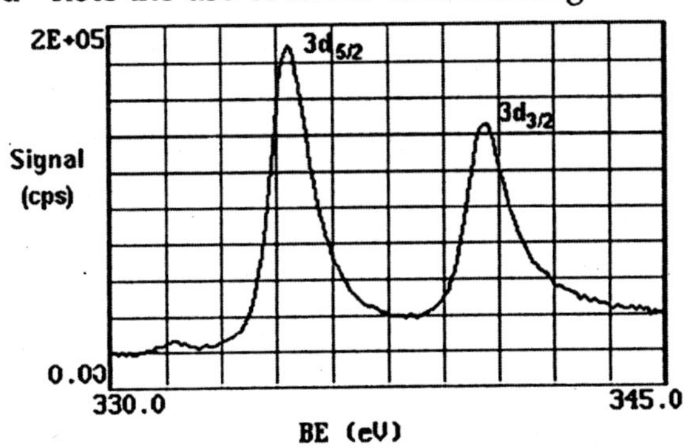


This spin-orbit splitting is of course not evident with *s*-levels ($l = 0$), but is seen with *p,d* & *f* core-levels which all show characteristic spin-orbit doublets.

## Chemical Shifts

The exact binding energy of an electron depends not only upon the level from which photoemission is occurring, but also upon:

1. The formal oxidation state of the atom
2. The local chemical and physical environment

Changes in either (1) or (2) give rise to small shifts in the peak positions in the spectrum - so-called *chemical shifts.*

Such shifts are readily observable and interpretable in XP spectra (unlike in Auger spectra) because the technique:

- Is of high intrinsic resolution (as core levels are discrete and generally of a well-defined energy)
- Is a one electron process (thus simplifying the interpretation)

Atoms of a higher positive oxidation state exhibit a higher binding energy due to the extra coulombic interaction between the photo-emitted electron and the ion core. This ability to discriminate between different oxidation states and chemical environments is one of the major strengths of the XPS technique.

In practice, the ability to resolve between atoms exhibiting slightly different chemical shifts is limited by the peak widths which are governed by a combination of factors; especially

- The intrinsic width of the initial level and the lifetime of the final state
- The line-width of the incident radiation - which for traditional x-ray sources can only be improved by using x-ray monochromators
- The resolving power of the electron-energy analyser

In most cases, the second factor is the major contribution to the overall line width.

## Angle Dependent Studies

The degree of surface sensitivity of an electron-based technique such as XPS may be varied by collecting photoelectrons emitted at different emission angles to the surface plane. This approach may be used to perform non-destructive analysis of the variation of surface composition with depth (with chemical state specificity).

**Ultraviolet Photoelectron Spectroscopy (UPS)**

In UPS the source of radiation is normally a noble gas discharge lamp; frequently a He-discharge lamp emitting He I radiation of energy 21.2 eV.

Such radiation is only capable of ionising electrons from the outermost levels of atoms - the valence levels. The advantage of using such UV radiation over x-rays is the very narrow line width of the radiation and the high flux of photons available from simple discharge sources. The main emphasis of work using UPS has been in studying:

1. The electronic structure of solids: detailed angle resolved studies permit the complete band structure to be mapped out in k-space.
2. The adsorption of relatively simple molecules on metals: by comparison of the molecular orbitals of the adsorbed species with those of both the isolated molecule and with calculations.

**Vibrational Spectroscopy**

Vibrational spectroscopy provides the most definitive means of identifying the surface species generated upon molecular adsorption and the species generated by surface reactions. In principle, any technique that can be used to obtain vibrational data from solid state or gas phase samples (IR, Raman etc.) can be applied to the study of surfaces - in addition there are a number of techniques which have been specifically developed to study the vibrations of molecules at interfaces (EELS, SFG etc.).

There are, however, only two techniques that are routinely used for vibrational studies of molecules on surfaces - these are:

1. IR Spectroscopy (of various forms, e.g. RAIRS, MIR)
2. Electron Energy Loss Spectroscopy (EELS)

## IR SPECTROSCOPY

There are a number of ways in which the IR technique may be implemented for the study of adsorbates on surfaces.

For solid samples possessing a high surface area:

Transmission IR Spectroscopy: employing the same basic experimental geometry as that used for liquid samples and mulls. This is often used for studies on supported metal catalysts where the large metallic surface area permits a high concentration of adsorbed species to be sampled. The solid sample must, of course, be IR transparent over an appreciable wavelength range.

Diffuse Reflectance IR Spectroscopy (DRIFTS): in which the diffusely scattered IR radiation from a sample is collected, refocused and analysed. This modification of the IR technique can be employed with high surface area catalytic samples that are not sufficiently transparent to be studied in transmission.

For studies on low surface area samples (e.g. single crystals):

Reflection-Absorption IR Spectroscopy (RAIRS): where the IR beam is specularly reflected from the front face of a highly-reflective sample, such as a metal single crystal surface.

Multiple Internal Reflection Spectroscopy (MIR): in which the IR beam is passed through a thin, IR transmitting sample in a manner such that it alternately undergoes total internal reflection from the front and rear faces of the sample. At each reflection, some of the IR radiation may be absorbed by species adsorbed on the solid surface - hence the alternative name of Attenuated Total Reflection (ATR).

### RAIRS - The Study of Adsorbates on Metallic Surfaces by Reflection IR Spectroscopy

It can be shown theoretically that the best sensitivity for IR measurements on metallic surfaces is obtained using a grazing-incidence reflection of the IR light.

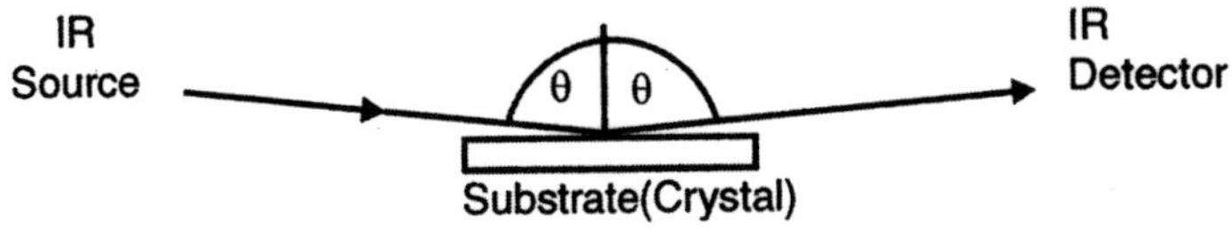


Furthermore, since it is an optical (photon in/photon out) technique it is not necessary for such studies to be carried out in vacuum. The technique is not inherently surface-specific, but

- There is no bulk signal to worry about
- The surface signal is readily distinguishable from gas-phase absorptions using polarization effects.

One major problem, is that of sensitivity (i.e. the signal is usually very weak owing to the small number of adsorbing molecules). Typically, the sampled area is ca. 1 $cm^2$ with less than $10^{15}$ adsorbed molecules (i.e. about 1 nanomole). With modern FTIR spectrometers, however, such small signals (0.01% – 2% absorption) can still be recorded at relatively high resolution (ca. 1 $cm^{-1}$).

For a number of practical reasons, low frequency modes (< 600 $cm^{-1}$) are not generally observable - this means that it is not usually possible to see the vibration of the metal-adsorbate bond and attention is instead concentrated on the intrinsic vibrations of the adsorbate species in the range 600 - 3600 $cm^{-1}$.

The observation of vibrational modes of adsorbates on metallic substrates is subject to the *surface dipole selection rule.*

This states that only those vibrational modes which give rise to an oscillating dipole perpendicular (normal) to the surface are IR active and give rise to an observable absorption band.

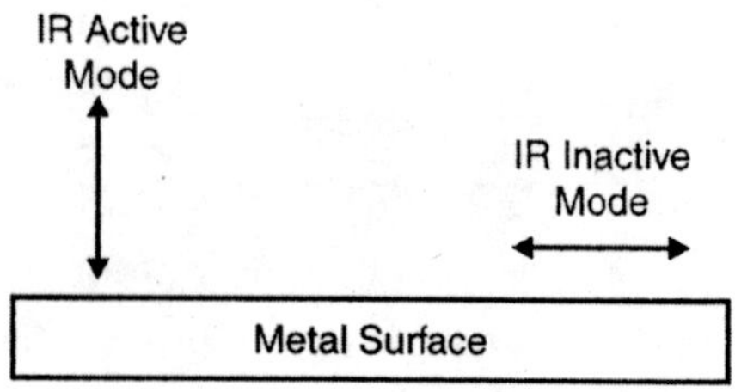


Further information on the selection rules for surface IR spectroscopy. It also needs to be remembered that even if a transition is allowed it may still be very weak if the transition moment is small.

## REFLECTION IR SPECTRA

Eg.1 Nitric oxide (NO) adsorption on a Pt surface.

Eg.2 HCN adsorption on a Pt surface

### Electron Energy Loss Spectroscopy (EELS)

This is a technique utilising the inelastic scattering of low energy electrons in order to measure vibrational spectra of surface species: superficially, it can be considered as the electron-analogue of Raman spectroscopy.

To avoid confusion with other electron energy loss techniques it is sometimes referred to as

## HREELS - HIGH RESOLUTION EELS *OR* VELS - VIBRATIONAL ELS

Since the technique employs low energy electrons, it is necessarily restricted to use in high vacuum (HV) and UHV environments - however, the use of such low energy electrons ensures that it is a surface specific technique and, arguably, it is the vibrational technique of choice for the study of most adsorbates on single crystal substrates.

The basic experimental geometry is fairly simple as illustrated schematically below - it involves using an electron monochromator to give a well-defined beam of electrons of a fixed incident energy, and then analysing the scattered electrons using an appropriate electron energy analyser.

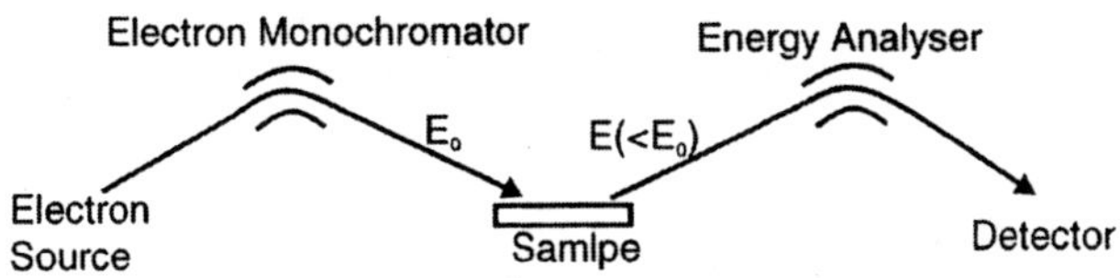


A substantial number of electrons are elastically scattered ($E = E_0$) - this gives rise to a strong *elastic peak* in the spectrum.

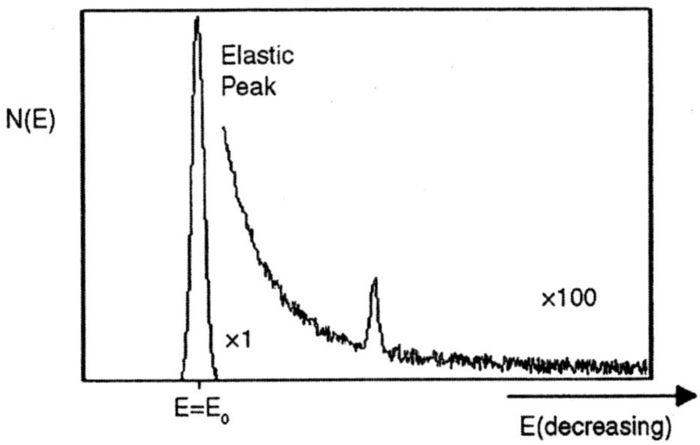


On the low kinetic energy side of this main peak ($E < E_0$), additional weak peaks are superimposed on a mildly sloping background. These peaks correspond to electrons which have undergone discrete energy losses during the scattering from the surface.

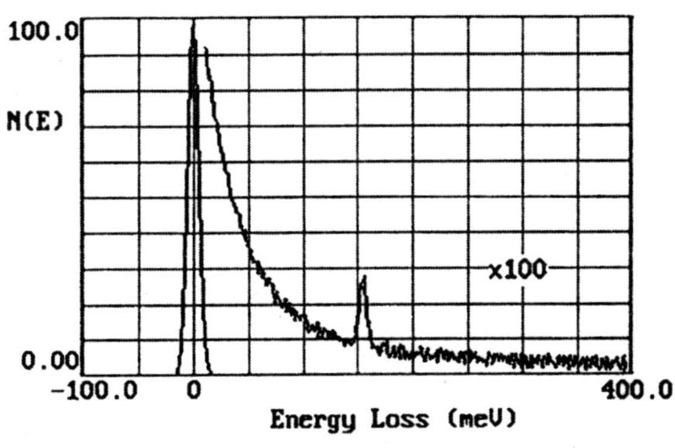


The magnitude of the energy loss, $\Delta E = (E_0 - E)$, is equal to the vibrational quantum (i.e. the energy) of the vibrational mode of the adsorbate excited in the inelastic scattering process. In practice, the incident energy ($E_0$) is usually in the range 5-10 eV (although occasionally up to 200 eV) and the data is normally plotted against the energy loss (frequently measured in meV).

The selection rules that determine whether a vibrational band may be observed depend upon the nature of the substrate and also the experimental geometry: specifically the angles of the incident and (analysed) scattered beams with respect to the surface.

For metallic substrates and a specular geometry, scattering is principally by a long-range dipole mechanism. In this case the loss features are relatively intense, but only those vibrations giving rise to a dipole change normal to the surface can be observed.

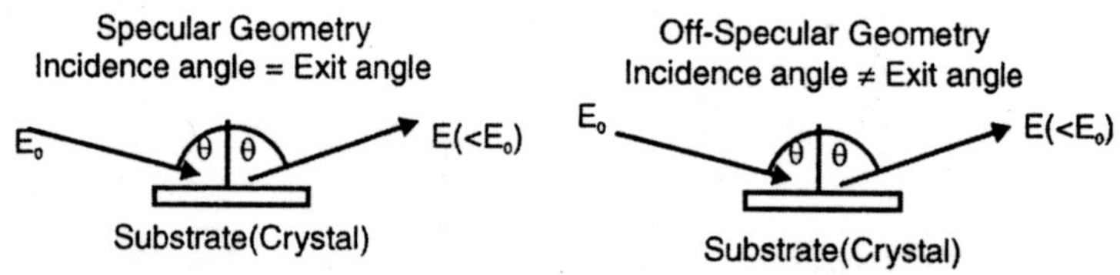


By contrast, in an off-specular geometry, electrons lose energy to surface species by a short-range impact scattering mechanism. In this case the loss features are relatively weak but all vibrations are allowed and may be observed.

If spectra can be recorded in *both* specular and off-specular modes the selection rules for metallic substrates can be put to good use - helping the investigator to obtain more definitive identification of the nature and geometry of the adsorbate species.

The resolution of the technique (despite the HREELS acronym !) is generally rather poor; 40-80 $cm^{-1}$ is not untypical. A measure of the instrumental resolution is given by looking at the FWHM (full-width at half maximum) of the elastic peak.

This poor resolution can cause problems in distinguishing between closely similar surface species - however, recent improvements in instrumentation have opened up the possibility of much better spectral resolution (< 10 $cm^{-1}$) and will undoubtedly enhance the utility of the technique.

In summary, there are both advantages and disadvantages in utilising EELS, as opposed to IR techniques, for the study of surface species It offers the advantages of...

- High sensitivity
- Variable selection rules
- Spectral acquisition to below 400 $cm^{-1}$

but suffers from the limitations of...

- Use of low energy electrons (requiring a HV environment and hence the need for low temperatures to study weakly-bound species, and also the use of magnetic shielding to reduce the magnetic field in the region of the sample)
- Requirement for flat, preferably conducting, substrates
- Lower resolution

## APPLICATIONS OF VIBRATIONAL SPECTROSCOPY

One of the classic examples of an area in which vibrational spectroscopy has contributed significantly to the understanding of the surface chemistry of an adsorbate is that of:

### Molecular Adsorption of CO on Metallic Surfaces

Adsorbed carbon monoxide usually gives rise to strong absorptions in both the IR and EELS spectra at the (C-O) stretching frequency. The metal-carbon stretching mode (ca. 400 $cm^{-1)}$ is usually also accessible to EELS.

The interpretation of spectra of CO as an adsorbed surface species draws heavily upon IR spectra from related inorganic cluster and coordination complexes - the structure of such complexes usually being available from x-ray single crystal diffraction measurements.

This comparison suggests that the CO stretching frequency can provide a good indication of the surface coordination of the molecule: as a rough guideline,

| | ν(C-O) |
|---|---|
| CO (gas phase) | 2143 $cm^{-1}$ |
| Terminal CO | 2100 – 1920 $cm^{-1}$ |
| Bridging (2f site) | 1920 – 1800 $cm^{-1}$ |
| Bridging (3f / 4f site) | < 1800 $cm^{-1}$ |

**RAIR spectrum for CO chemisorbed on a clean Pt surface**

The reduction in the stretching frequency of terminally-bound CO from the value observed for the gas phase molecule (2143 $cm^{-1}$) can be explained in terms of the Dewar-Chatt or *Blyholder model* for the bonding of CO to metals.

This simple model considers the metal-CO bonding to consist of two main components:

A

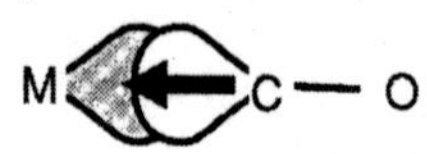


σ←pσ

B

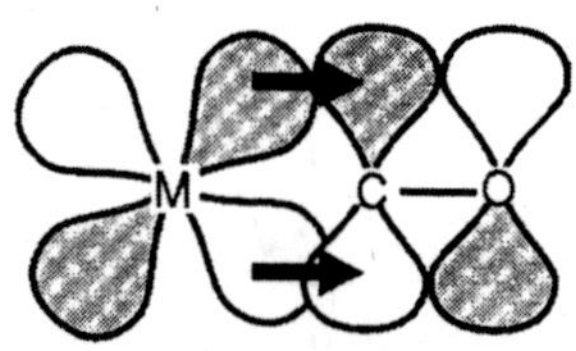


dπ→pπ

A: this is a σ bonding interaction due to overlap of a filled σ -"lone pair" orbital on the carbon atom with empty metal orbitals of the correct symmetry - this leads to electron density transfer from the CO molecule to the metal centre.

B: this is a π bonding interaction due to overlap of filled metal dπ (and pπ ?) orbitals with the π* antibonding molecular orbital of the CO molecule. Since this interaction leads to the introduction of electron density into the CO antibonding orbital there is a consequent reduction in the CO bond strength and its intrinsic vibrational frequency (relative to the isolated molecule).

## SECONDARY ION MASS SPECTROMETRY

The technique of Secondary Ion Mass Spectrometry (SIMS) is the most sensitive of all the commonly-employed surface analytical techniques this is because of the inherent sensitivity associated with mass spectrometric-based techniques.

There are a number of different variants of the technique:

- Statics SIMS: used for sub-monolayer elemental analysis
- Dynamic SIMS: used for obtaining compositional information as a function of depth below the surface
- Imaging SIMS: used for spatially-resolved elemental analysis

All of these variations on the technique are based on the same basic physical process and it is this process, together with a brief introduction to the field of static SIMS. Further notes on dynamic and imaging SIMS can be obtained.

In SIMS the surface of the sample is subjected to bombardment by high energy ions - this leads to the ejection (or *sputtering*) of both neutral and charged (+/-) species from the surface. The ejected species may include atoms, clusters of atoms and molecular fragments.

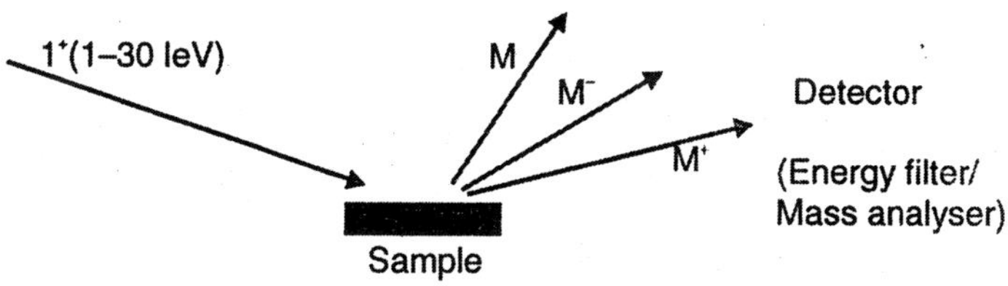


In traditional SIMS it is only the positive ions that are mass analysed - this is primarily for practical ease but it does lead to problems with quantifying the compositional data since the positive ions are but a small, non-representative fraction of the total sputtered species. It should be further noted that the displaced ions have to be energy filtered before they are mass analysed (i.e. only ions with kinetic energies within a limited range are mass analysed).

The most commonly employed incident ions (denoted by $I^+$ in the above diagram) used for bombarding the sample are argon ions ($Ar^+$) but other ions (e.g. alkali metal ions, $Ga^+$) are preferred for some applications.

The mass analyser is typically a quadrupole MS analyser with unit mass resolution, but high specification time-of-flight (TOF) analysers are also used and provide substantially higher sensitivity and a much greater mass range (albeit at a higher cost).

### Static SIMS

In static SIMS (SSIMS) the aim is to obtain sufficient signal to provide compositional analysis of the surface layer without actually removing a significant fraction of a monolayer - i.e. to be able to analyse less than ca. $10^{14}$ atoms/molecules (about 10% of a monolayer for a $1 cm^2$ sample). The technique is then capable of providing information about the topmost single atomic layer of the surface.

*Example*: static SIMS spectra from the surface of PTFE (polytetrafluoroethylene)

Note the substantial differences between the positive and negative ion spectra.

In each case the most intense peaks correspond to relatively stable positive and negative ions respectively.

## TEMPERATURE-PROGRAMMED TECHNIQUES

There are a range of techniques for studying surface reactions and molecular adsorption on surfaces which utilise temperature-programming to discriminate between processes with different activation parameters. Of these, the most useful for single crystal studies is:

### Temperature Programmed Desorption (TPD)

When the technique is applied to a system in which the adsorption process is, at least in part, irreversible and *T*-programming leads to surface reactions, then this technique is often known as:

### Temperature Programmed Reaction Spectroscopy (TPRS)

However, there is no substantive difference between TPRS and TPD.

The basic experiment is very simple, involving

1. Adsorption of one or more molecular species onto the sample surface at low temperature (frequently 300 K, but sometimes sub-ambient).

2. Heating of the sample in a controlled manner (preferably so as to give a linear temperature ramp) whilst monitoring the evolution of species from the surface back into the gas phase.

In modern implementations of the technique the detector of choice is a small, quadrupole mass spectrometer (QMS) and the whole process is carried out under computer control with quasi-simultaneous monitoring of a large number of possible products.

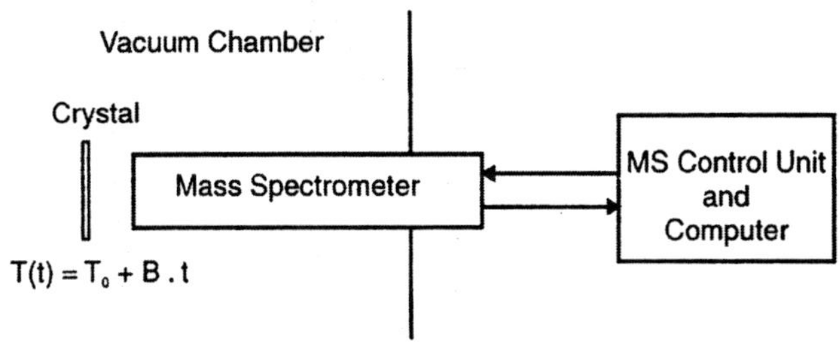


The data obtained from such an experiment consists of the intensity variation of each recorded mass fragment as a function of time / temperature. In the case of a simple reversible adsorption process it may only be necessary to record one signal - that attributable to the molecular ion of the adsorbate concerned.

The graph below shows data from a TPD experiment following adsorption of CO onto a Pd(111) crystal at 300 K.

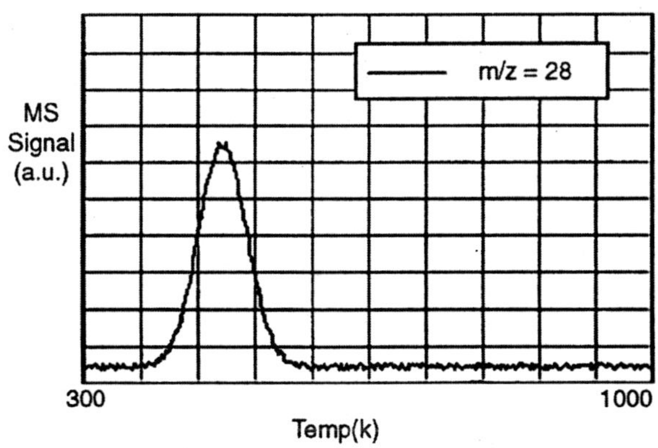

Since mass spectrometric detection is used the sensitivity of the technique is good with attainable detection limits below 0.1% of a monolayer of adsorbate.

The following points are worth noting:

1. The area under a peak is proportional to the amount originally adsorbed, i.e. proportional to the surface coverage.
2. The kinetics of desorption (obtained from the peak profile and the coverage dependence of the desorption characteristics) give information on the state of aggregation of the adsorbed species e.g. molecular v's dissociative.
3. The position of the peak (the *peak temperature*) is related to the enthalpy of adsorption, i.e. to the strength of binding to the surface.

One implication of the last point, is that if there is more than one binding state for a molecule on a surface (and these have significantly different adsorption enthalpies) then this will give rise to multiple peaks in the TPD spectrum.

The graph below shows data from a TPD experiment following adsorption of oxygen on Pt(111) at 80 K.

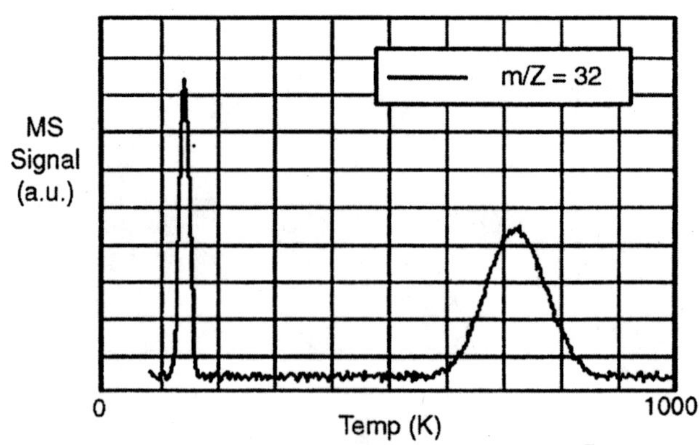


## THEORY OF TPD

The rate of desorption of a surface species will in general be given by an expression of the form:

$R_{des} = n.\ N^x \exp(-E_a^{des} / R\ T)$

where $R_{des}$ - desorption rate (= $-dN/dt$)

$x$– kinetic order of desorption (typically 0,1 or 2)

$E_a^{des}$ - activation energy for desorption

In a temperature programmed desorption experiment in which the temperature is increased linearly with time from some initial temperature $T_o$, then:

$T = T_o + \beta.t$ and $dT = \beta.dt$

The intensity of the desorption signal, $I(T)$, is proportional to the rate at which the surface concentration of adsorbed species is decreasing. This is obtained by combining to give

$$I(T)\alpha - \frac{dN}{dT} = \frac{VN^x}{\beta}\exp\left(\frac{-E_a^{des}}{RT}\right)$$

This problem may also be considered in a rather simplistic graphical way -the key to this is to recognise that the expression for the desorption signal given in the above equation is basically a product of a coverage term ($N^x$ - where $N$ depends on $T$) and an exponential term (involving both $E_a$ and $T$).

Initially, at low temperatures $E_a >> RT$ and the exponential term is vanishingly small. However, as the temperature is increased this term begins to increase very rapidly when the value of $RT$ approaches that of the activation energy, $E_a$.

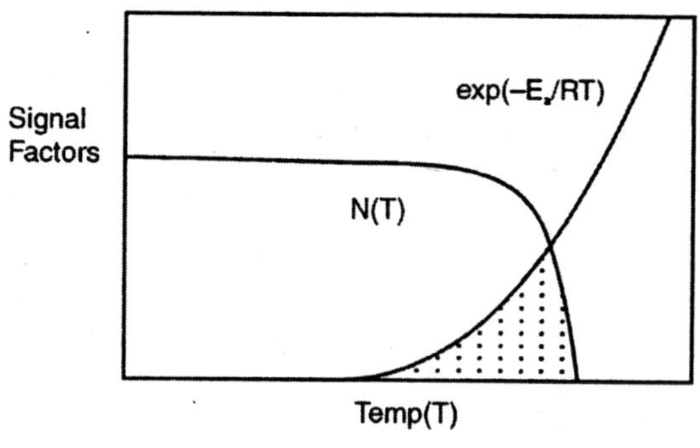

By contrast, the pre-exponential term is dependent upon the coverage, $N(T)$, at the temperature concerned - this term will remain at the initial value until the desorption rate becomes of significance as a result of the increasing exponential term. Thereafter, it will decrease ever more rapidly until the coverage is reduced to zero. The shaded area is an approximate representation of the product of these two functions, and hence also an approximate representation for the desorption signal itself - whilst this illustration may be overly simplistic it does clearly show why the desorption process gives rise to a well-defined desorption peak.

CASE I (Molecular adsorption) - in this case the desorption kinetics will usually be first order (i.e. $x = 1$)

The maximum desorption signal in the $I(T)$ trace will occur when $(dI / dT) = 0$,

i.e. when

$$\frac{d}{dT}\left[\frac{VN}{\beta}\exp\left(\frac{-E_a^{des}}{RT}\right)\right] = 0$$

Hence, remembering that the surface coverage changes with temperature i.e.

$$N = N(T), \frac{VN}{\beta}\cdot\frac{E_a}{RT^2}\exp\left(\frac{-E_a}{RT}\right) + \frac{v}{\beta}\exp\left(\frac{-E_a}{RT}\right)\cdot\frac{dN}{dT}$$

$= 0$

where we have substituted $E_a$ for $E_a^{des}$ purely for clarity of presentation.

Substituting for $dN/dT$ from eqtn.then gives

$$\frac{vN}{\beta}\left[\cdot\frac{E_a}{RT^2} - \frac{v}{\beta}\exp\left(\frac{-E_a}{RT}\right)\right]\exp\left(\frac{-E_a}{RT}\right)$$

$= 0$

The solution is given by setting the expression in square brackets to be equal to zero, i.e.

$$\frac{E_a}{RT_p^2} = \frac{\nu}{\beta}\exp\left(\frac{-E_a}{RT_p}\right)$$

where we have now defined the temperature at which the desorption maximum occurs to be $T = T_p$ (the *peak temperature*).

Unfortunately, this equation cannot be re-arranged to make $T_p$ the subject (i.e. to give a simple expression of the form $T_p$ =.....), but we can note that:

1. As $E_a^{des}$ (the activation energy for desorption) increases, then so $T_p$ (the peak temperature) increases.
2. The peak temperature is not dependent upon, and consequently does not change with, the initial coverage, $N_{t=0}$.
3. The shape of the desorption peak will tend to be asymmetric, with the signal decreasing rapidly after the desorption maximum.

## TEMPERATURE PROGRAMMED REACTION SPECTROSCOPY

In TPRS a number of desorption products will normally be detected in any one experiment - this is where mass spectrometric detection and multiple ion monitoring really becomes essential.

A good example of TPRS results are those obtained from dosing formic acid (HCOOH) onto a copper surface.

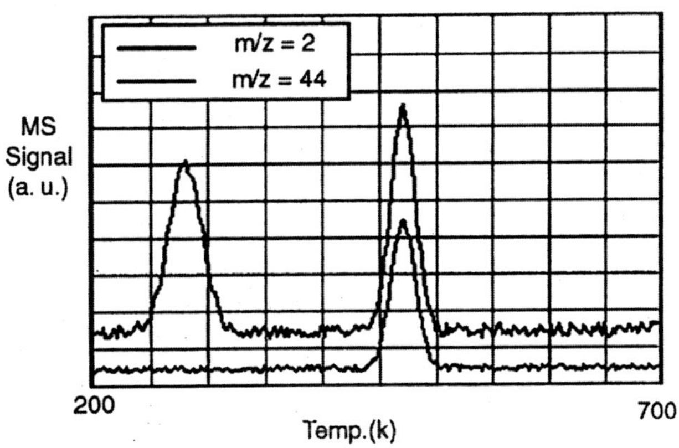

The spectra contain two regions of interest - the first at around room temperature where there is desorption of hydrogen ($m/z = 2$) alone, and the second between 400 & 500 K where there are coincident desorption peaks attributable to hydrogen and carbon dioxide ($m/z = 44$).

The lower temperature hydrogen peak is at about the same temperature at which hydrogen atoms recombine and desorb as molecular hydrogen from a hydrogen-dosed Cu(110) surface i.e. the kinetics of its evolution into the gas phase are those of the normal recombinative desorption process. By contrast, the higher temperature hydrogen peak is observed well above the normal hydrogen desorption temperature - its appearance must be governed by the kinetics of decomposition of another surface species. The carbon dioxide peak must also be *decomposition limited* since on clean Cu(110) carbon dioxide itself is only weakly physisorbed at very low temperatures.

More generally, you should note that coincident desorption of two species at temperatures well above their normal desorption temperatures is characteristic of the decomposition of a more complex surface species.

# 4

# Surface Imaging and Depth Profiling

## BASIC CONCEPTS IN SURFACE IMAGING AND LOCALISED SPECTROSCOPY

Most surface spectroscopic techniques involve probing the surface by exposing it to a flux of "particles" (hn, $e^-$, $A^+$....) and simultaneously monitoring the response to this stimulation by, for example, measuring the energy distribution of emitted electrons. In their most basic form, these techniques collect information from a relatively large area of surface (~ $mm^{2)}$. In most cases, however, there are variations of these techniques which permit either,

1. Information to be collected only from a specific region of the surface - that is.... Localised Spectroscopy

   or

2. The spatial distribution of a property to be mapped - that is Surface Microscopy

The requirement in both cases is for spatial localisation of the spectroscopic technique. This may be achieved in one of two ways,

### Localisation of the Probe

Spatial localisation of a spectroscopic technique can be achieved by focusing the probe so that it interacts with only a small region of the surface.

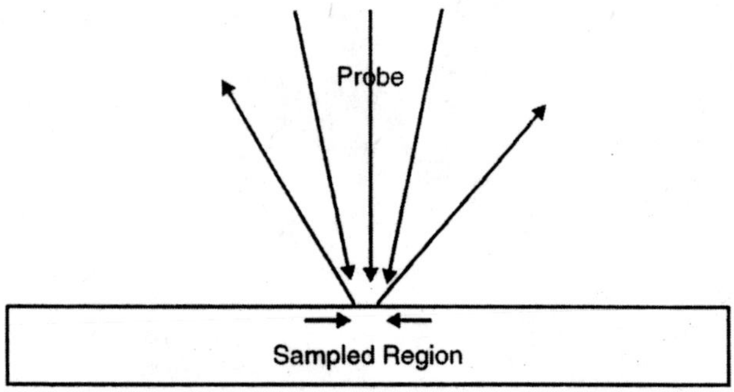


The ease with which this can be accomplished and the spatial resolution (degree of localisation) attainable depends primarily upon the nature of the probe - charged particles (e.g. electrons and ions) are easily focused; neutral particles (e.g. x-ray photons and neutral atoms) are not readily focused.

With this localisation of the probe into a small area, it is clearly advantageous to collect as many as possible of the outgoing particles whose detection completes the spectroscopic process in order to maximize the signal-to-noise.

In the diagram above, the probe is stationary and incident upon a particular small region of the surface in a manner suitable for localised surface spectroscopy. In order to carry out surface imaging or mapping, either

- The probe must be scanned (*rastered*) across the surface, or
- The surface must be moved under the probe beam.

**Spatially-localized Detection**

The alternative approach is to allow a large area of the sample to be exposed to the probe but only to collect signal (i.e. emitted particles) from a small region of the sample. Thus, in this instance, the spatial localisation is occurring on the detection side of the technique.

In principle, this is easy - it simply requires an appropriate lens system to collect as much emission as possible from the desired part of the surface and focus it onto a detector as illustrated in a highly schematic fashion below.

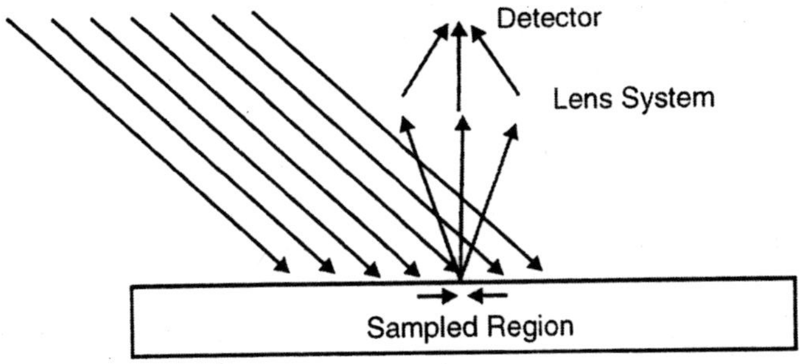


In practice, this approach has only really been successfully employed in surface analysis when the emitted particles are electrons.

In order to get an image of the surface using a single detector it is necessary to either

1. Scan the sample position underneath the detection system.
2. Make use of additional scanning plates in the electon-optical focusing system.

An alternative approach to obtaining images using this same basic arrangement involves employing a linear array or two-dimensional array of detecting elements. In this type of system, the emission from different parts of the surface is effectively focussed onto different parts of the array detector as schematically illustrated below.

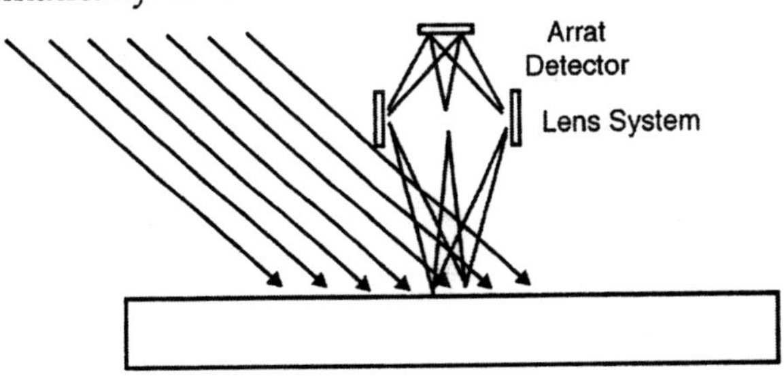


Some additional scanning mechanism is still required to build-up an image or map of anything other than the directly imaged region of surface, but the use of multiple (array) detectors greatly enhances the speed of image acquisition and ultimately the image quality.

In summary, spatial localisation of a surface spectroscopic technique may be obtained by either focusing the probe or imaging only a small part of the sample surface onto a detector using a lens system. Images and compositional maps are then usually obtained by scanning the focused probe in the first case, and either translating the sample or employing sophisticated optical/detector systems in the latter case.

## ELECTRON MICROSCOPY - SEM AND SAM

The two forms of electron microscopy which are commonly used to provide surface information are

*Secondary Electron Microscopy (SEM)*: which provides a direct image of the topographical nature of the surface from all the emitted secondary electrons

*Scanning Auger Microscopy (SAM)*: which provides compositional maps of a surface by forming an image from the Auger electrons emitted by a particular element.

Both techniques employ focusing of the probe beam (a beam of high energy electrons, typically 10 - 50 keV in energy) to obtain spatial localisation.

### Secondary Electron Microscopy (SEM)

As the primary electron beam is scanned across the surface, electrons of a wide range of energies will be emitted from the surface in the region where the beam is incident. These electrons will include backscattered primary electrons and Auger electrons, but the vast majority will be *secondary electrons* formed in multiple inelastic scattering processes (these are the electrons that contribute to the background and are completely ignored in Auger spectroscopy). The secondary electron current reaching the detector is recorded and the microscope image consists of a "plot" of this current, I, against probe position on the surface. The contrast in the micrograph arises from several mechanisms, but first and foremost from variations in the surface topography. Consequently, the secondary electron micrograph is virtually a direct image of the real surface structure.

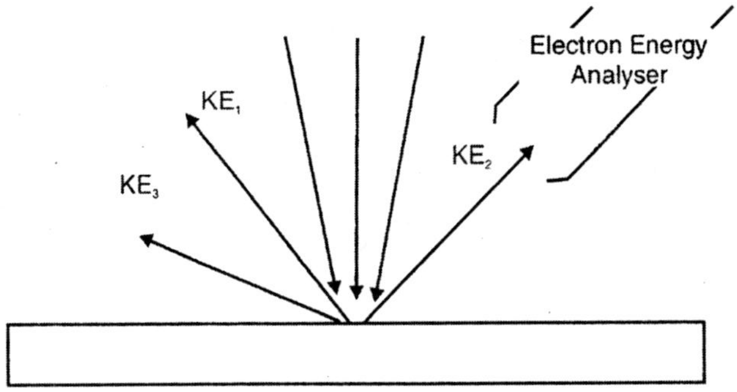


The attainable resolution of the technique is limited by the minimum spot size that can be obtained with the incident electron beam, and ultimately by the scattering of this beam as it interacts with the substrate. With modern instruments, a resolution of better than 5 nm is achievable. This is more than adequate for imaging semiconductor device structures, for example, but insufficient to enable many supported metal catalysts to be studied in any detail.

## SCANNING AUGER MICROSCOPY (SAM)

The incident primary electrons cause ionization of atoms within the region illuminated by the focused beam. Subsequent relaxation of the ionized atoms leads to the emission of Auger electrons characteristic of the elements present in this part of the sample surface.

As with SEM, the attainable resolution is again ultimately limited by the incident beam characteristics. More significantly, however, the resolution is also limited by the need to acquire sufficient Auger signal to form a respectable image within a reasonable time period, and for this reason the instrumental resolution achievable is rarely better than about 15-20 nm.

## IMAGING XPS

The combination of the features of X-ray photoelectron spectroscopy (in particular, quantitative surface elemental analysis and chemical state information) together with spatial localisation is a particularly desirable option in surface analysis. However, whilst much progress has been made in developing the technique of imaging XPS, there is still a considerable research effort being devoted to improving the available spatial resolution beyond that which is currently available.

Different manufacturers of imaging XPS systems have adopted different strategies for obtaining spatial localisation - including all of those mentioned in. Specificially, these include

1. Localisation of the probe, by focusing the incident x-rays.
2. Localisation by selected (limited) area analysis.
3. Use of array detectors, with associated imaging optics.

## IMAGING XPS BY X-RAY FOCUSSING

Until very recently this was not a popular option in commercial instruments, since x-rays are rather difficult to focus (in fact x-ray lens design is a science in its own right and alternative focusing technique, usually employing specially-bent diffracting crystals, have to be employed).

Nevertheless, with recent improvements in technology, an x-ray spot size of better than 10 mm has been achieved using this approach; images may then be obtained *either* by scanning the sample under the focused x-ray beam, *or* by scanning the microfocused x-ray beam.

## LIMITED AREA ANALYSIS

The most common approach to localisation of analysis, is to restrict the area of the sample surface from which photoelectrons are analysed using a combination of lenses and apertures in the design of the electron energy analyser.

The main problem with this approach is that as the sampled area is reduced, so is the collected signal - consequently, there is a direct trade-off between spatial resolution and data collection time.

The practically achievable spatial resolution is rarely better than 100 mm.

Imaging of the sample surface may then be achieved by either

1. Translating the sample position under the electron energy analyser, so that the analysed region is moved across the surface.
2. Incorporating electrostatic deflection plates within the electron optics to move the region from which electrons are collected across the sample surface.

One advantage of this technique is its relative simplicity (and hence relatively low cost) - but this is reflected in the poor performance !

**Array Detectors and Imaging Optics**

There are a number of innovative designs for imaging spectrometers which combine array (i.e. multiple-segment) detectors with sophisticated imaging optics to obtain electron-energy resolved images at much faster acquisition rates. The spatial resolution achievable using this approach is also higher than that for either of the other techniques mentioned - state-of-the-art instruments give better than 5 mm resolution

**State of the Art**

This is one area in which an ultra-high intensity x-ray source offers major advantages and the highest performance imaging XPS systems (scanning photoemission microscopes) are therefore those based at synchrotron sources. Using zone-plate technology to focus the x-rays (combined with multi-segment detectors to enhance data acquisition rates) it is possible on such systems to obtain XPS images with a resolution of better than 100 nm.

In summary, the spatial resolution currently achievable with commercial imaging XPS instruments limits their potential applications - nevertheless there are many areas of materials science where the information obtainable is incredibly useful and the relatively poor spatial resolution (compared, for example, with electron microscopic techniques such as SAM) is more than offset by the benefit of concurrent chemical state definition.

## SIMS - IMAGING AND DEPTH PROFILING

The basic ideas behind the SIMS (Secondary Ion Mass Spectrometry) technique. The technique utilises a beam of atomic ions (charged particles) as the probe, it is a relatively easy matter to focus the incident beam and then to scan it across the surface to give an imaging technique.

## SURFACE IMAGING USING SIMS

If the aim of the measurement is to obtain compositional images of the surface formed from the secondary ion spectrum with minimum possible damage to the surface, then the main problem is to ensure that sufficient signal is obtained at the desired spatial resolution whilst minimizing the ion flux incident on any part of the surface.

This is most easily achieved by switching from the traditional instrumental approach of using continuous-flux ion guns and quadrupole mass spectrometer detectors, to using pulsed ion sources and time-of-flight (TOF) mass spectrometers. The TOF mass spectrometers are a much more efficient way of acquiring spectral data, and also provide good resolution and sensitivity up to very high masses. Using such instruments, SIMS images with a spatial resolution of better than 50 nm are obtainable.

### SIMS Depth Profiling

The aim of depth profiling is to obtain information on the variation of composition with depth below the initial surface - such information is obviously particularly useful for the analysis of layered structures such as those produced in the semiconductor industry.

Since the SIMS technique itself relies upon the removal of atoms from the surface, it is by its very nature a destructive technique, but also, ideally suited for depth profiling applications. Thus a depth profile of a sample may be obtained simply by recording sequential SIMS spectra as the surface is gradually eroded away by the incident ion beam probe. A plot of the intensity of a given mass signal as a function of time, is a direct reflection of the variation of its abundance/ concentration with depth below the surface.

One of the main advantages that SIMS offers over other depth profiling techniques (e.g. Auger depth profiling) is its sensitivity to very low concentrations (~ ppm) of elements - again this is particularly important in the semiconductor industry where dopants are often present at very low concentrations.

The *depth resolution* achievable (the ability to discriminate between atoms in adjacent thin layers) is dependent upon a number of factors which include:

1. The uniformity of etching by the incident ion beam
2. The absolute depth below the original surface to which etching has already been carried out
3. The nature of the ion beam utilized (i.e. the mass & energy of the ions)

as well as effects related to the physics of the sputtering process itself (e.g. ion-impact induced burial).

## AUGER DEPTH PROFILING

Auger Spectroscopy is a surface sensitive spectroscopic technique yielding compositional information.

In its basic form it provides compositional information on a relatively large area (~ 1 $mm^{2)}$ of surface using a broad-focused electron beam probe. In this manner, sufficient signal can be readily obtained whilst keeping the incident electron flux low, and thus avoiding potential electron-induced modifications of the surface. As a consequence the technique is non-destructive when used in this manner.

To obtain information about the variation of composition with depth below the surface of a sample, it is necessary to gradually remove material from the surface region being analysed, whilst continuing to monitor and record the Auger spectra.

This controlled surface etching of the analysed region can be accomplished by simultaneously exposing the surface to an ion flux which leads to sputtering (i.e. removal) of the surface atoms.

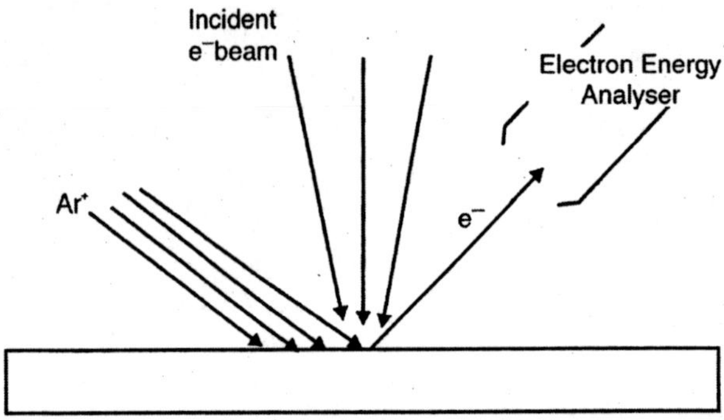


For example, suppose there is a buried layer of a different composition several nanometres below the sample surface.

As the ion beam etches away material from the surface, the Auger signals corresponding to the elements present in this layer will rise and then decrease again.

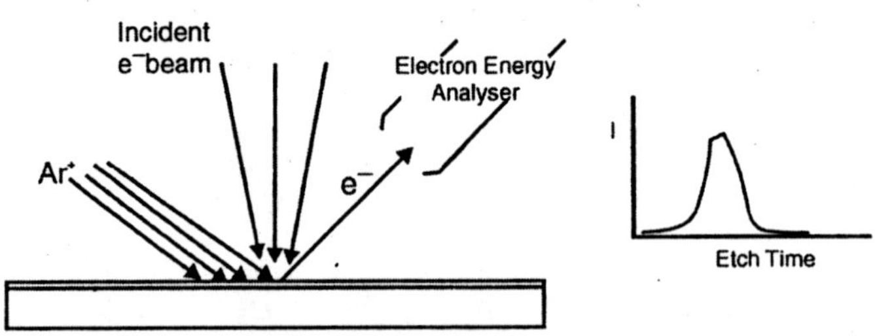


The diagram above shows the variation of the Auger signal intensity one might expect from such a system for an element that is only present in the buried layer and not in the rest of the solid.

In summary, by collecting Auger spectra as the sample is simultaneously subjected to etching by ion bombardment, it is possible to obtain information on the variation of composition with depth below the surface - this technique is known by the name of *Auger Depth Profiling*

## SCANNING PROBE MICROSCOPY (STM / AFM)

In the early 1980's two IBM scientists, developed a new technique for studying surface structure - Scanning Tunnelling Microscopy (STM). This invention was quickly followed by the development of a whole family of related techniques which, together with STM, may be classified in the general category of Scanning Probe Microscopy (SPM) techniques. Of these later techniques, the most important is Atomic Force Microscopy (AFM).

The development of these techniques has without doubt been the most important event in the surface science field in recent times, and opened up many new areas of science and engineering at the atomic and molecular level.

## BASIC PRINCIPLES OF SPM TECHNIQUES

All of the techniques are based upon scanning a probe (typically called the *tip* in STM, since it literally is a sharp metallic tip) just above a surface whilst monitoring some interaction between the probe and the surface.

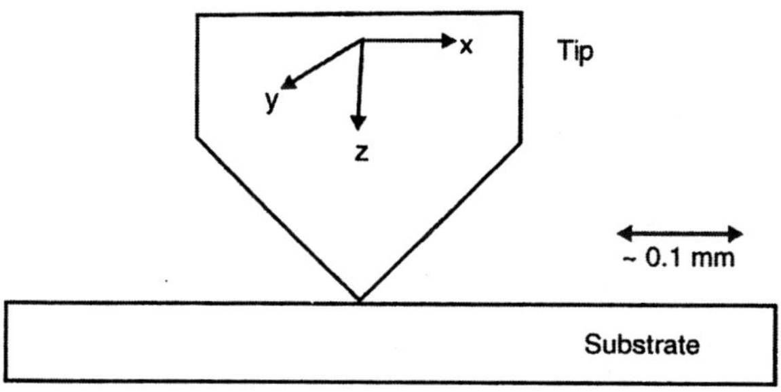


The interaction that is monitored in:

STM - is the *tunnelling current* between a metallic tip and a conducting substrate which are in very close proximity but not actually in physical contact.

AFM - is the van der Waals force between the tip and the surface; this may be either the short range repulsive force (in contact-mode) or the longer range attractive force (in non-contact mode).

For the techniques to provide information on the surface structure at the atomic level (which is what they are capable of doing !):

1. The position of the tip with respect to the surface must be very accurately controlled (to within about 0.1 Å) by moving either the surface or the tip.
2. The tip must be very sharp - ideally terminating in just a single atom at its closest point of approach to the surface.

The attention paid to the first problem and the engineering solution to it is the difference between a good microscope and a not so good microscope - it need not worry us here, sufficient to say that it is possible to accurately control the relative positions of tip and surface by ensuring good vibrational isolation of the microscope and using sensitive piezoelectric positioning devices.

Tip preparation is a science in itself - having said that, it is largely serendipity which ensures that one atom on the tip is closer to the surface than all others.

Let us look at the region where the tip approaches the surface in greater detail.

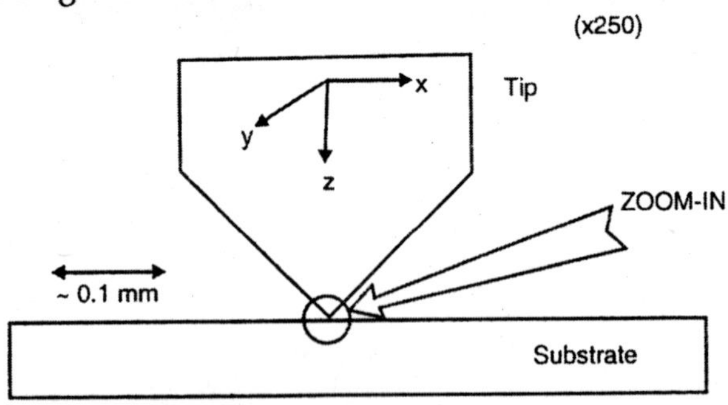


The end of the tip will almost invariably show a certain amount of structure, with a variety of crystal facets exposed.

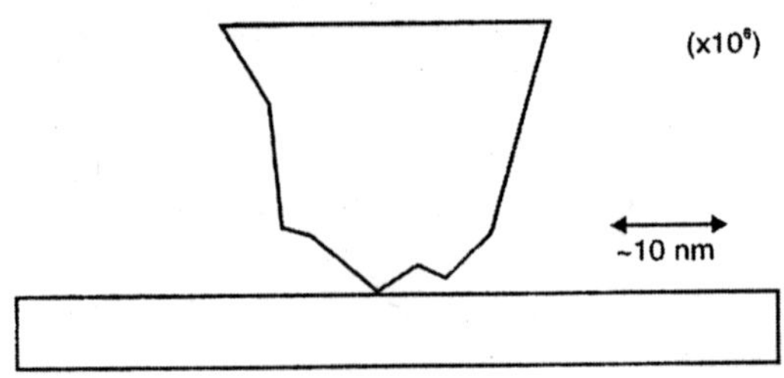

and if we now go down to the atomic scale.

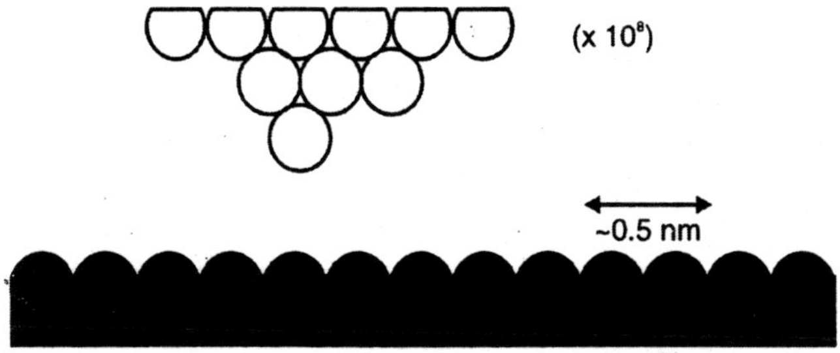


there is a reasonable probability of ending up with a truly atomic tip.

If the tip is biased with respect to the surface by the application of a voltage between them then electrons can tunnel between the two, provided the separation of the tip and surface is sufficiently small - this gives rise to a *tunnelling current*.

The direction of current flow is determined by the polarity of the bias.

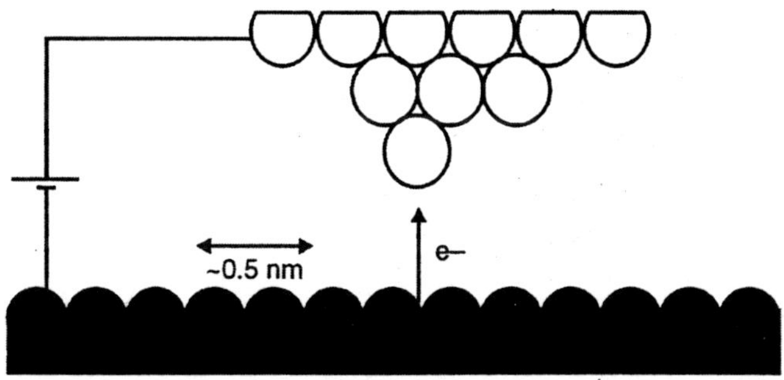


If the sample is biased -ve with respect to the tip, then electrons will flow from the surface to the tip as shown above, whilst if the sample is biased +ve with respect to the tip, then electrons will flow from the tip to the surface as shown below.

The name of the technique arises from the quantum mechanical *tunnelling*-type mechanism by which the electrons can move between the tip and substrate. Quantum mechanical tunnelling permits particles to tunnel through a potential barrier which they could not surmount according to the classical laws of physics - in this case electrons are able to traverse the classically-forbidden region between the two solids as illustrated schematically on the energy diagram below.

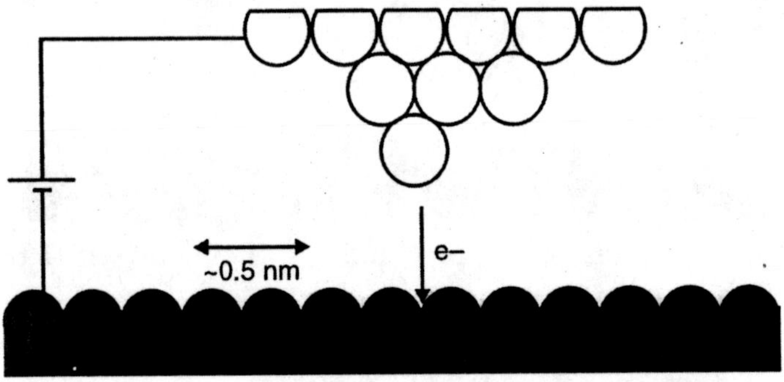


This is an over-simplistic model of the tunnelling that occurs in STM but it is a useful starting point for understanding how the technique works.

In this model, the probability of tunnelling is exponentially-dependent upon the distance of separation between the tip and surface: the tunnelling current is therefore a very sensitive probe of this separation.

Imaging of the surface topology may then be carried out in one of two ways:

1. In constant height mode (in which the tunnelling current is monitored as the tip is scanned parallel to the surface)
2. In constant current mode (in which the tunnelling current is maintained constant as the tip is scanned across the surface)

If the tip is scanned at what is nominally a constant height above the surface, then there is actually a periodic variation in the separation distance between the tip and surface atoms. At one point the tip will be directly above a surface atom and the tunnelling current will be large whilst at other points the tip will be above hollow sites on the surface and the tunnelling current will be much smaller.

A plot of the tunnelling current v's tip position therefore shows a periodic variation which matches that of the surface structure - hence it provides a direct "image" of the surface (and by the time the data has been processed it may even look like a real picture of the surface !).

In practice, however, the normal way of imaging the surface is to maintain the tunnelling current constant whilst the tip is scanned across the surface. This is achieved by adjusting the tip's height above the surface so that the tunnelling current does not vary with the lateral tip position. In this mode the tip will move slightly upwards as it passes over a surface atom, and conversely, slightly in towards the surface as it passes over a hollow.

The image is then formed by plotting the tip height (strictly, the voltage applied to the z-piezo) v's the lateral tip position.

The development of the various scanning probe microscopy techniques has revolutionized the study of surface structure - atomic resolution images have been obtained not only on single crystal substrates in UHV but also on samples at atmospheric pressure and even under solution. Many problems still remain, however, and the interpretation of SPM data is not always as straightforward as it might at first seem. There is still very much a place for the more traditional surface structural techniques such as LEED.

This introduction to STM has concentrated on the non-invasive imaging applications of the technique, yet there is increasing interest in using such techniques as a tool for the actual modification of surfaces. At the moment this is still at the "gimmicky" stage, but the longer term implications of being able to manipulate surface structure and molecules at the atomic level have yet to be fully appreciated: we can but await the future with interest.

# 5

# Surface Tension

Surface tension is a measurement of the cohesive energy present at an interface. The molecules of a liquid attract each other. The interactions of a molecule in the bulk of a liquid are balanced by an equal attractive force in all directions. Molecules on the surface of a liquid experience an imbalance of forces as indicated below.

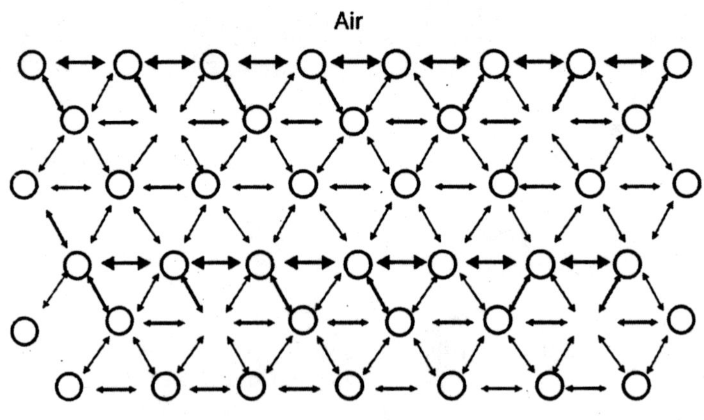


The net effect of this situation is the presence of free energy at the surface. The excess energy is called surface free energy and can be quantified as a measurement of energy/area. It is also possible to describe this situation as having a line tension or surface tension which is quantified as a force/length measurement. The common units for surface tension are dynes/cm or mN/m. These units are equivalent. This excess energy exists at the interface of two fluids. If one of the fluids is the vapour phase of a liquid being tested the measurement is referred to as surface tension. If the surface investigated is the interface of two liquids the measurement is referred to as interfacial tension. In either case the more dense fluid is referred to herein as the 'heavy phase' and the less dense fluid is referred to as the 'light phase'. Solids also may be described to have a surface free energy at their interfaces but direct measurement of its value is not possible through techniques used for liquids.

Polar liquids, such as water, have strong intermolecular interactions and thus high surface tensions. Any factor which decreases the strength of this interaction will lower surface tension. Thus an increase in the temperature of this system will lower surface tension. Any contamination, especially by surfactants, will lower surface tension. Therefore researchers should be very cautious about the issue of contamination.

## TENSIOMETRY

The measurement of surface and interfacial tension as performed by a tensiometer is based on force measurements of the interaction of a probe with the surface of interface of two fluids. With any of the techniques described herein you may perform interfacial tension measurements just like surface tension measurements by insuring that the bulk of the probe is submersed in the light phase prior to beginning the experiment.

In these experiments a probe is hung on a balance and brought into contact with the liquid interface tested. The forces experienced by the balance as the probe interacts with the surface of the liquid can be used to calculate surface tension. The forces present in this situation depend on the following factors; size and shape of the probe, contact angle of the liquid/solid interaction and surface tension of the liquid. The size and shape of the probe are easily controlled. The contact angle is controlled to be zero (complete wetting). This is achieved by using probes with high energy surfaces. KSV probes are made of a platinum/iridium alloy which insures complete wetting and easy and reliable cleaning.

The mathematical interpretation of the force measurements depends on the shape of the probe used. Two types of probes are commonly used, the DuNouy Ring and the Wilhelmy Plate. Both are available from KSV Instruments.

## CHOICE OF PROBES

DuNouy ring: This method utilizes the interaction of a platinum ring with the surface being tested. The ring is submerged below the interface and subsequently raised upwards. As the ring moves upwards it raises a meniscus of the liquid. Eventually this meniscus tears from the ring and returns to it's original position. Prior to this event, the volume, and thus the force exerted, of the meniscus passes through a maximum value and begins to diminish prior to the actually tearing event. The process is shown in the diagram below:

1. The ring is above the surface and the force is zeroed.
2. The ring hits the surface and there is a slight positive force because of the adhesive force between the ring and the surface.
3. The ring must be pushed through the surface (due to the surface tension) which causes a small negative force.
4. The ring breaks through the surface and a small positive force is measured due to the supporting wires of the ring.
5. When lifted through the surface the measured force starts to increase.
6. The force keeps increasing until
7. The maximum force is reached
8. After the maximum there is a small decrease of in the force until the lamella breaks.

The calculation of surface or interfacial tension by this technique is based on the measurement of this maximum force.

The depth of immersion of the ring and the level to which it is raised when it experiences the maximum pull are irrelevant to this technique. The original calculations based on the ring technique were based on theories which apply to rings of infinite diameter and do not consider an additional volume of liquid which is raised due to the proximity of one side of the ring to the other. This additional liquid lifted is diagrammed below as the shaded portion:

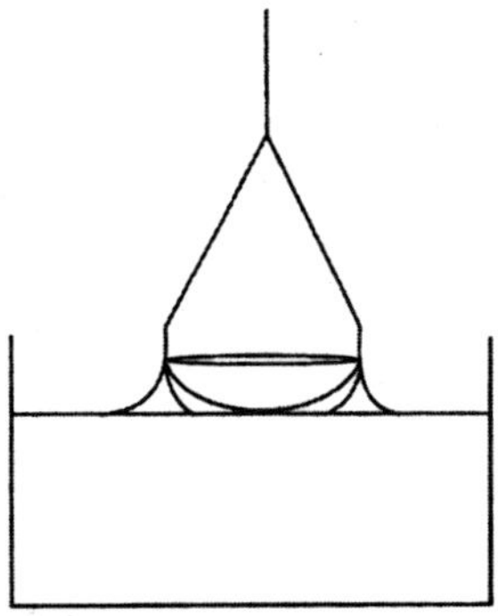

Mathematical corrections which compensate for this extra liquid have been produced.

Wilhelmy Plate: This method utilizes the interaction of a platinum plate with the surface being tested. The calculations for this technique are based on the geometry of a fully wetted plate in contact with, but not submerged in, the heavy phase. In this method the position of the probe relative to the surface is significant. As the surface is brought into contact with the probe your Sigma 70 will notice this event by the change in forces it experiences. It will register the height at which this occurs as the 'zero depth of immersion'. The plate will then be wetted to a set depth to insure that there is indeed complete wetting of the plate(zero contact angle). When the plate is later returned to the zero depth of immersion, the force it registers can be used to calculate surface tension.

Metal rod: If you wish to measure surface tension of a liquid of which you have limited quantities you may consider using a platinum rod as your probe. The liquid required for testing may be more than you have available. The amount necessary to test with a certain probe must be sufficient to fill a vessel which accommodates the probe,to a depth which insures complete wetting of the probe is possible. With a given volume you can fill to a greater depth if the diameter of the vessel is smaller. Limitations to this strategy occur when the edges of the probe approach the edges of the measuring vessel. In that case the forces on the balance may be effected by the meniscus of the liquid as it contacts the vessel.

To avoid these problems you can choose to employ a metal rod as your probe. This allows you to use very much smaller volumes of your test liquids. In any technique the absolute accuracy of the measurement of the geometry of your probe effects the accuracy of your results. Measurement of the geometry of a fine rod are likely to be less accurate, on a percentage basis, than the measurement of your larger probes. Therefore this approach should only be used in situations where the volume of liquid available is an issue. The calculations for this technique are based on the same principles as the Wilhelmy plate method.

## PENDANT DROP SHAPE ANALYSIS

The shape of a drop of liquid hanging from a syringe tip is determined from the balance of forces which include the surface tension of that liquid. The surface or interfacial tension at the liquid interface can be related to the drop shape through the following equation:

$$\gamma = \Delta\rho\, g\, R_0^2 / \beta$$

where $\gamma$ = surface tension

$\Delta\rho$ = difference in density between fluids at interface

$g$ = gravitational constant

$R_0$ = radius of drop curvature at apex

$\beta$ = shape factor

$\beta$, the shape factor can be defined through the Young-Laplace equation expressed as 3 dimensionless first order equations.

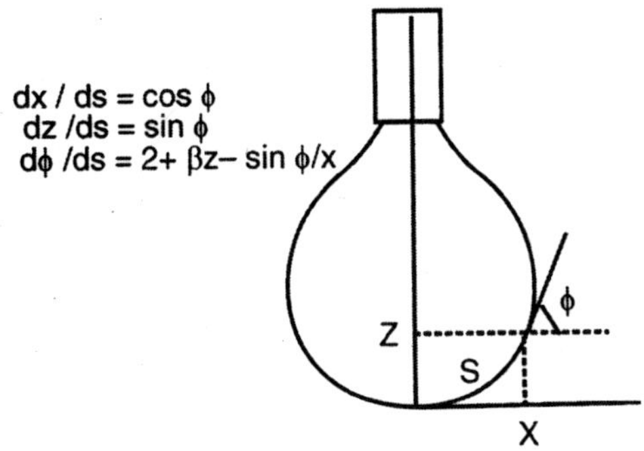


Modern computational methods using iterative approximations allow solution of the Young-Laplace equation for $\beta$ to be performed. Thus for any pendant drop where the densities of the two fluids in contact are known, the surface tension may be measured based upon the Young-Laplace equation. This approach represents a significant improvement, in both ease and accuracy, from traditional methods.

## CONTACT ANGLE

Contact angle,$\theta$, is a quantitative measure of the wetting of a solid by a liquid. It is defined geometrically as the angle formed by a liquid at the three phase boundary where a liquid, gas and solid intersect as shown below:

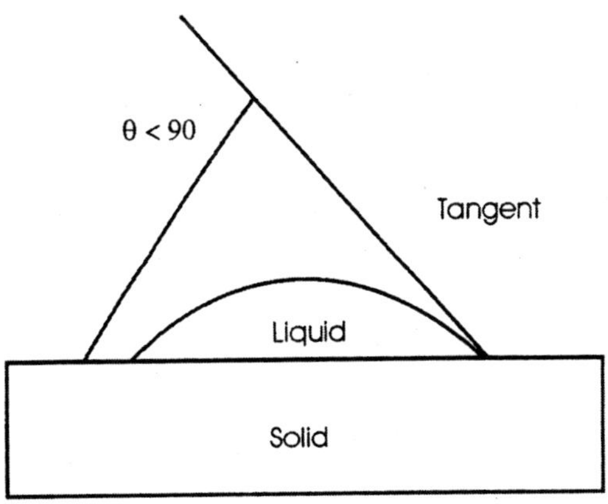


It can be seen from this figure that low values of $\theta$ indicate that the liquid spreads, or wets well, while high values indicate poor wetting. If the angle q is less than 90 the liquid is said to wet the solid. If it is greater than 90 it is said to be non-wetting. A zero contact angle represents complete wetting.

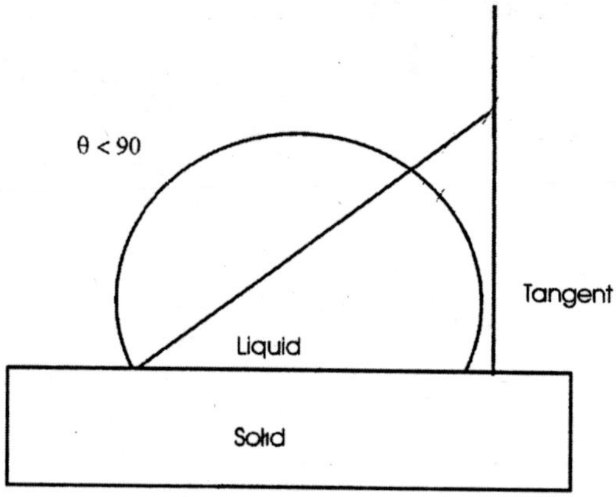


The measurement of a single static contact angle to characterize the interaction is no longer thought to be adequate. For any given solid/ liquid interaction there exists a range of contact angles which may be found. The value of static contact angles are found to depend on the recent history of the interaction. When the drop has recently expanded the angle is said to represent the 'advanced' contact angle. When the drop has recently contracted the angle is said to represent the 'receded' contact angle. These angles fall within a range with advanced angles approaching a maximum value and receded angles approaching a minimum value.

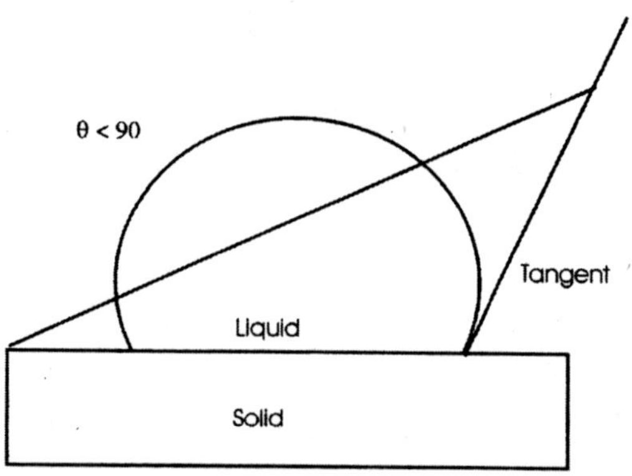

If the three phase(liquid/solid/vapour) boundary is in actual motion the angles produced are called Dynamic Contact Angles and are referred to as 'advancing' and 'receding' angles. The difference between 'advanced' and 'advancing', 'receded' and 'receding' is that in the static case motion is incipient in the dynamic case motion is actual. Dynamic contact angles may be assayed at various rates of speed. Dynamic contact angles measured at low velocities should be equal to properly measured static angles.

## HYSTERESIS

The difference between the maximum(advanced/advancing) and minimum(receded/receding) contact angle values is called the contact angle hysteresis. A great deal of research has gone into analysis of the significance of hysteresis. It has been used to help characterize surface heterogeneity, roughness and mobility. Briefly, for surfaces which are not homogeneous there will exist domains on the surface which present barriers to the motion of the contact line. For the case of chemical heterogeneity these domains represent areas with different contact angles than the surrounding surface. For example when wetting with water, hydrophobic domains will pin the motion of the contact line as the liquid advances thus increasing the contact angles. When the water recedes the hydrophilic domains will hold back the draining motion of the contact line thus decreasing the contact angle. From this analysis it can be seen that, when testing with water, advancing angles will be sensitive to the hydrophobic domains and receding angles will characterize the hydrophilic domains on the surface.

For situations in which surface roughness generates hysteresis the actual microscopic variations of slope in the surface create the barriers which pin the motion of the contact line and alter the macroscopic contact angles. There has been a great deal of research investigating the significance of hysteresis and you are recommended to the papers cited at the end of this note for further details.

Contact angle can also be considered in terms of the thermodynamics of the materials involved. This analysis involves the interfacial free energies between the three phases and is given by:

$$\gamma_{lv} \cos \theta = \gamma_{sv} - \gamma_{sl}$$

where $\gamma_{lv}$, $\gamma_{sv}$ and $\gamma_{sl}$ refer to the interfacial energies of the liquid/vapour, solid/vapour and solid/liquid interfaces.

**How is Contact Angle Measured?**

Two different approaches are commonly used to measure contact angles of non-porous solids, goniometry and tensiometry. Goniometry involves the observation of a sessile drop of test liquid on a solid substrate. Tensiometry involves measuring the forces of interaction as a solid is contacted with a test liquid. Both techniques are described below with comments on the choice of either technique for particular research applications.

**GONIOMETRY**

Analysis of the shape of a drop of test liquid placed on a solid is the basis for goniometry. The basic elements of a goniometer include a light source, sample stage, lens and image capture. Contact angle can be assessed directly by measuring the angle formed between the solid and the tangent to the drop surface.

The production of drops with advanced and receded edges involves one of two strategies. Drops can be made to have advanced edges by addition of liquid. Receded edges may be produced by allowing sufficient evapouration or by withdrawing liquid from the drop. Alternately, both advanced and receded edges are produced when the stage on which the solid is held is tilted to the point of incipient motion. Using an instrument with high speed image capture capabilities shapes of drops in motion may be analysed.

**ADVANTAGES**

Goniometry can be used in many situations where tensiometry cannot. You can use a great variety of solid substrates provided they have a relatively flat portion for testing and can fit on the stage of the instrument. Substrates with regular curvature, such as contact lenses are also easily analysed.Testing can be done using very small quantities of liquid. It is also easy to test high temperature liquids such as polymer melts.

## LIMITATIONS

The assignment of the tangent line which will define the contact angle is a factor which can limit the reproducibility of contact angle measurements. Conventional goniometry relies on the consistency of the operator in the assignment of the tangent line. This can lead to significant error, especially subjective error between multiple users.

The conditions which produce advanced and receded angles are sometimes difficult to reproduce. Although drops in motion can produce data on dynamic contact angles the velocity of motion cannot be controlled. It is also less suited, when compared to tensiometry, to analysis of the effects of wetting on changes in contact angle.

In addition the amount of surface sampled for each measurement is limited and multiple measurements should be used to characterize a surface. Fibres are not easily studied by goniometry.

## TENSIOMETRY

The tensiometric method for measuring contact angles measures the forces that are present when a sample of solid is brought into contact with a test liquid. If the forces of interaction, geometry of the solid and surface tension of the liquid are known the contact angle may be calculated. The user first makes a measurement of the surface tension of the liquid using either a Wilhelmy plate or DuNouy ring. The sample of the solid to be tested is then hung on the balance and tared. The liquid is then raised to contact the solid. When the solid contacts the liquid the change in forces is detected and your Sigma70 registers this elevation as zero depth of immersion.

As the solid is pushed into the liquid the forces on the balance are recorded. The forces on the balance are

$F_{total}$ = wetting force + weight of probe – buoyancy

Your Sigma70 has tared the weight of the probe and can remove the effects of the buoyancy force by extrapolating the graph back to zero depth of immersion. The remaining component force is the wetting force which is defined as:

Wetting force = $\gamma_{LV} P \cos\theta$

where $\gamma_{LV}$ is the liquid surface tension, P is the perimeter of the probe and $\theta$ is the contact angle. Thus at any depth data is received which can be used to calculate contact angle. This contact angle, which is obtained from data generated as the probe advances into the liquid, is the advancing contact angle. The sample is immersed to a set depth and the process is reversed. As the probe retreats from the liquid data collected is used to calculate the receding contact angle.

**Advantages**

The use of tensiometry for measurement of contact angle has several advantages over conventional goniometry. At any point on the immersion graph, all points along the perimeter of the solid at that depth contribute to the force measurement recorded. Thus the force used to calculate $\theta$ at any given depth of immersion is already an averaged value. You may calculate an averaged value for the entire length of the sample or average any part of the immersion graph data to assay changes in contact angle along the length of the sample.

This technique allows the user to analyse contact angles produced from wetting over an entire range of velocities from static to rapid wetting. Because the contact angles are determined from the forces measured by the instrument there is no possibility of subjective error.

The graphs produced by this technique are very useful in studying hysteresis. Variations of contact angles, both advancing and receding, for the entire length of the sample tested are visualized on the same graph. In addition variations generated over multiple wetting/dewetting cycles can yield information on changes caused by wetting(such as absorption or surface reorientation).

Analysis of fibres, very problematic for goniometry, is handled easily by your tensiometer.

**Limitations**

There are two major limitations for the application of this technique. Firstly the user must have enough of the liquid being tested available so that he can immerse a portion of his solid in it. Secondly the solid in question must be available in samples which meet the following constraints. The sample must be formed or cut in a regular geometry such that it has a constant perimeter over a portion of it's length. Rods, plates or fibres of known perimeter are ideal. The sample must have the same surface on all sides which contact the liquid.

It is also more difficult to use this technique in systems which are measured at high temperatures. Temperatures at or below 100 ° C are easily handled but for measurements above this range goniometry is recommended.

**Washburn Method**

This method is chosen when the solid sample to be tested contains a porous architecture which leads to absorption of the wetting liquid. The solid is brought into contact with the testing liquid and the mass of liquid absorbed into the solid is measured as a function of time. The amount absorbed is a function of the viscosity, density and surface tension of the liquid, the material constant of the solid, and the contact angle of the interaction. If the viscosity, density and surface tension of the liquid are known the material constant and contact angle can be solved for.

## UTILIZATION OF CONTACT ANGLE DATA

The primary focus of contact angle studies is in assessing the wetting characteristics of solid/liquid interactions. Contact angle is commonly used as the most direct measure of wetting.

Other experimental parameters may be derived directly from contact angle and surface tension results. Some examples are:

Work of Adhesion: defined as the work required to separate the liquid and solid phases, or the negative free energy associated with the adhesion of the solid and liquid phases. Used to express the strength of the interaction between the two phases. It is given by the Young-Dupre equation as:

$$W_a = \gamma (1 + \cos \theta)$$

Work of Cohesion: defined as the work required to separate a liquid into two parts, it is a measure of the strength of molecular interactions within the liquid. It is given by;

$$W_c = 2 \gamma$$

Work of Spreading: the negative free energy associated with spreading liquid over solid surface. Also referred to as Spreading Coefficient it is given as:

$$W_s = \gamma (\cos \theta - 1)$$

Wetting Tension: a measurement of force/length defined as:

$$\tau = F_w / P = \gamma_{LV} \cos \theta$$

This value, wetting force normalized for length, also represents the product of the cosine of the contact angle and the surface tension. It allows for a characterization of the strength of the wetting interaction without separate measurement of surface tension. Most helpful in situations, such as multicomponent systems, where surface tension at interface may not equal equilibrium surface tension. Also referred to as Adhesion Tension or Work of Wetting.

## CHARACTERIZATION OF THE SOLID SURFACE

Measurements of surface tension yield data which directly reflect thermodynamic characteristics of the liquid tested. Measurement of contact angles yield data which reflect the thermodynamics of a liquid/solid interaction. If you wish to characterize the wetting behaviour of a particular liquid/solid pair you only need to report the contact angle. It is possible to characterize the wettability of your solid in a more general way. Various methods are used but the same basic principle applies for each. The solid is tested against a series of liquids and contact angles are measured. Calculations based on these measurements produce a parameter(critical surface tension, surface free energy,etc) which quantifies a characteristic of the solid which mediates wetting. Two basic approaches are covered here

Critical Surface Tension: Using a series of homologous liquids of differing surface tensions a graph of $\cos\theta$ vs $\gamma$ is produced. It will be found that the data form a line which approaches $\cos\theta = 1$ at a given value of $\gamma$. This is the maximal surface tension of a liquid which may completely wet your solid. This value, called the critical surface tension, can be used to characterize your solid surface. See references for details on procedure and limitations.

Free Surface Energy: Another way to characterize a solid surface is by calculating free surface energy, also referred to as solid surface tension. This approach involves testing the solid against a series of well characterized wetting liquids. The liquids used must be characterized such that the polar and dispersive components of their surface tensions are known. The relevant equation is given by Owens and Wendt as:

$$\gamma_l (1+ \cos \theta)/(\gamma_{ld})^{1/2} = (\gamma_{sp})^{1/2} [(\gamma_{lp})^{1/2} /(\gamma_{ld})^{1/2}] + (\gamma_{sd})^{1/2}$$

where $\theta$ is the contact angle, $\gamma_l$ is liquid surface tension and $\gamma_s$ is the solid surface tension, or free energy. The addition of d and p in the subscripts refer to the dispersive and polar components of each. The form of the equation is of the type $y = mx + b$. You can graph $(\gamma_{lp})^{1/2} /(\gamma_{ld})^{1/2}$ vs $\gamma_l$ (1+ cos q) / $(g_{ld})^{1/2}$. The slope will be $(g_{sp})^{1/2}$ and the y-intercept will be $(g_{sd})^{1/2}$. The total free surface energy is merely the sum of its two component forces.

The tensiometric method for measuring contact angles measures the forces that are present when a sample of solid is brought into contact with a test liquid. If the forces of interaction, geometry of the solid and surface tension of the liquid are known the contact angle may be calculated. The user first makes a measurement of the surface tension of the liquid using either a Wilhelmy plate or DuNouy ring. The sample of the solid to be tested is hung on the balance and tared. The liquid is then raised to contact the solid. When the solid contacts the liquid the change in forces is detected and your Sigma70 registers this elevation as zero depth of immersion. As the solid is pushed into the liquid the forces on the balance are recorded. The forces on the balance are

$F_{total}$ = wetting force + weight of probe - buoyancy

Your Sigma70 has tared the weight of the probe and can remove the effects of the buoyancy force by extrapolating the graph back to zero depth of immersion. The remaining component force is the wetting force which is defined as:

Wetting force = $g_{LV}$ P cosq

where $\gamma_{LV}$ is the liquid surface tension, P is the perimeter of the probe and $\theta$ is the contact angle. Thus at any depth data is received which can be used to calculate contact angle. This contact angle, which is obtained from data generated as the probe advances into the liquid, is called the 'advancing contact angle'. The sample is immersed to a set depth and the process is reversed. As the probe retreats from the liquid data collected is used to calculate a 'receding contact angle'. The process will appear as follows:

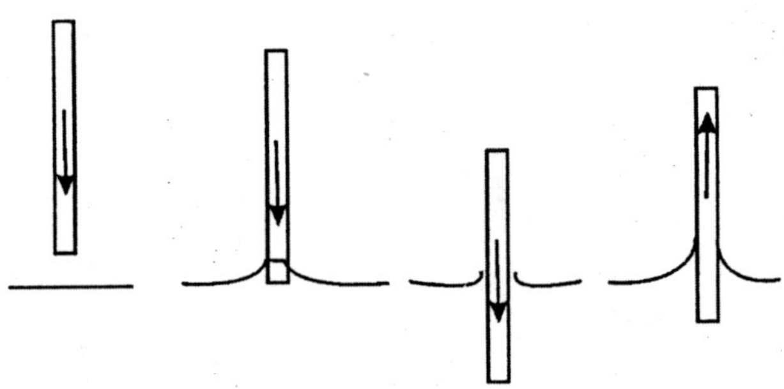

Advancing Receding

1. The sample is above the liquid and the force/ length is zeroed.
2. The sample hits the surface. For the sample as shown, with a contact angle < 90°, the liquid rises up.... causing a positive force.

3. The sample is immersed, buoyant force increases causing a decrease in force on the balance. Forces are measured for advancing angle.
4. After having reached the desired depth the sample is pulled out of the liquid. Forces are measured for receding angle.

The contact angles measured in this way are termed Dynamic Contact Angles. The difference between static advanced/receded contact angles and advancing/receding dynamic contact angles is that in the static case motion is incipient while in the dynamic case motion is actual. Dynamic contact angles may be assayed at various rates of speed. Dynamic contact angles measured at low velocities should be equal to properly measured static angles.

The use of tensiometry to measure contact angles was outlined quite early in the literature of wetting measurements. However, general application of this technique was significantly delayed due to the absence of commercially available tensiometers which had the precision and versatility needed for easy measurement of contact angles. Without a sophisticated tensiometer designed for contact angle measurement the Wilhelmy plate technique is difficult. The Sigma 70 is a precision tensiometer integrated with a lifting mechanism and controlled through a PC which now makes measurements of contact angles fast and accurate.

**Advantages**

The use of tensiometry for measurement of contact angle has several advantages over conventional goniometry. At any point on the immersion graph, all points along the perimeter of the solid at that depth contribute to the force measurement recorded. Thus the force used to calculate $\theta$ at any given depth of immersion is already an averaged value. You may calculate an averaged value for the entire length of the sample or average any part of the immersion graph data to assay changes in contact angle along the length of the sample. Hysteresis, the difference between advancing and receding angles, is very easy to determine using this method. Using multiple cycles you can identify variations in surface structure and notice adsorption phenomena. Analysis of fibres, very problematic for goniometry, is handled easily by your tensiometer. Fibres with diameters below 10 microns should give reliable data. It is also easy to test the contact angle of coatings. Simply coat a solid substrate, for example a microscope slide, and test the coated solid.

The references for this note include several papers comparing the efficacy of tensiometry and goniometry for measuring contact angles.

### Limitations

There are two major limitations for the application of this technique. Firstly the user must have enough of the liquid being tested available so that he can immerse a portion of his solid in it. Secondly the solid in question must be available in samples which meet the following constraints. The sample must be formed or cut in a regular geometry such that it has a constant perimeter over a portion of it's length. Rods, plates or fibres of known perimeter are ideal. The sample must have the same surface on all sides which contact the liquid. The sample must also be small enough so that it can be hung on the microbalance of your tensiometer.

## POWDER WETTABILITY

Wetting studies usually involve measurement of contact angles as primary data. Contact anglese indicate the degree of wetting when a solid and liquid interact. The lower the contact angle the greater the wetting. Contact angles below 90° indicate that the liquid wets the solid spontaneously (in a thermodynamic rather than kinetic sense). Contact angles above 90° indicate that the liquid does not wet the solid.

If you wish to characterize the wetting behaviour of a particular liquid/solid pair you only need to measure the contact angle. It is possible to characterize the wettability of your solid in a more general way. Various methods are used but the same basic principle applies for each. The solid is tested against a series of liquids and contact angles are measured. Calculations based on these measurements produce a parameter (critical surface tension, surface free energy, etc) which quantifies a characteristic of the solid which mediates wetting.

The wetting of powders and porous solids also involves contact angle phenomena but is complicated by the presence of a porous architecture. Although various methods are employed to characterize this architecture, contact angles are still the primary parameters used to characterize wetting.

Rough estimates of wettability can be made by placing powders on any given liquid and observing if the solid wets into the liquid (the time required for wetting may also be noted). Solids with contact angles above 90° will not wet into the liquid. A series of liquids of varying surface tensions can be tested against your solid. Liquids with lower surface tensions yield lower contact angles. For some liquids the contact angle with a given solid will be greater than 90° and the solid will not wet while with a liquid of lower surface tension the contact angle will be less than 90° and wetting will occur. Using this crude approach some relevant information on wetting of solids may be found. To produce a quantitative measure of wetting, a contact angle, two approaches are generally used: goniometry and tensiometry.

### Goniometry

The goniometric approach to contact angles involves placing a drop of liquid on a flat solid surface and analyzing the shape of the drop formed. With either a porous solid or powder packed to approximate a flat surface, penetration of the liquid into the pores or spaces between particles may occur. If the real contact angle between the liquid and the solid is less than 90° 'spontaneous' penetration will occur. Spontaneous, as used in this context, refers to the thermodynamic favourability of the event. The actual rate of penetration may vary considerably. In some cases it may occur so slowly that reasonable values of contact angle may be measured. In other cases the penetration occurs quite rapidly. Researchers interested in studying the dynamics of wetting in this situation can use KSV's CAM 200 goniometer to capture changing contact angles at high speed. The CAM 200 is capable of recording at 40 msec intervals and contains on on screen trigger. This exclusive feature allows the researcher to begin the recording based on the occurrence of specific events, such as contact of liquid with the solid surface. Using this approach the researcher can produce data on the changes in apparent contact angle vs time to quantify the wetting behaviour.

If the real contact angle of the interaction is greater than 90° penetration should not occur and reproducible contact angles may be measured.

### Tensiometry

The Sigma70 tensiometer may be used to measure contact angles of powders and porous solids. The Wilhelmy method for measuring contact angles has significant limitations when applied to powders. Some strategies include binding the powder to a solid substrate and proceeding to measure contact angle on the sample produced. Interactions with the binding element(eg tape) and quantification of the amount bound introduce factors which may effect results. An alternate approach is therefore recommended. This approach applies the Washburn theory of wetting of porous solids.

### Washburn Method

According to this theory when a porous solid is brought into contact with a liquid the rise of the liquid into the pores of the solid will obey the following relationship:

$$T = [\eta / C\, r^2\, \gamma \cos\theta]\, M^2$$

The terms are defined as follows:

$T$ = time after contact

$\eta$ = viscosity of liquid

$C$ = material constant characteristic of solid sample

$\rho$ = density of liquid

$\gamma$ = surface tension of liquid

$\theta$ = contact angle

M = mass of liquid adsorbed on solid

An experiment is performed where the mass of the adsorbed liquid is measured vs time. A graph of Time vs Mass$^2$ should yield a straight line whose slope is $h / C\rho^2 \gamma \cos\theta$. If viscosity, density and surface tension (found using the same tensiometer)are known from separate experiments there are only two unknowns left in this term, contact angle and C the material constant for the solid. To resolve this situation an experiment is performed in which the contact angle may be assumed to be zero. This should occur when a liquid with very low surface tension is used. N-hexane would be an appropriate choice. When such an experiment is performed, the material constant for the solid may be solved for.[slope = $\eta / C r^2 \gamma$]. It is then possible to perform the same experiment with any test liquid and calculate the contact angle for that system. Washburn experiments thus involve testing your solid with a completely wetting liquid, assuming $\theta = 0$, and solving for C. Then, assuming C is constant, you test the same solid against various liquids to find contact angles.

Factors effecting C: The logic of this method only holds in situations where C can be assumed to remain constant for a given solid from one experiment to the next. C, the material constant, is a term which reflects the porosity of the solid and the packing of particles(in the case of powders). Because this value is *assumed* to be constant in the Washburn method any actual variation will generate error in the calculation of contact angles. It is therefore critical to attempt to insure that C actually is constant.

Porous Solids The material constant C reflects the number and orientation of the pores present in the solid sample. If you are testing porous solids that retain a definite shape(i.e. paper, fabric, etc.) you need only insure that each sample has the same size, shape and orientation.

Packing For testing powders the material constant will depend on the porosity of individual particles and the packing of the particles. If all samples of your powder are from the same source the porosity may be reasonably assumed to be constant. The major variations in actual C values are usually due to variations in the packing of powders. Various methods have been attempted to try to assure consistent packing and you are welcome to try any technique. Most researcher use a technique which involves using a reproducible force to pack a known mass of powder into a sleeve with a porous bottom(either sintered glass or screen lined with filter paper). KSV is currently using vibration on a modified audio speaker to control packing. The interaction of the sleeve containing the powder and the liquid interface is diagrammed below.

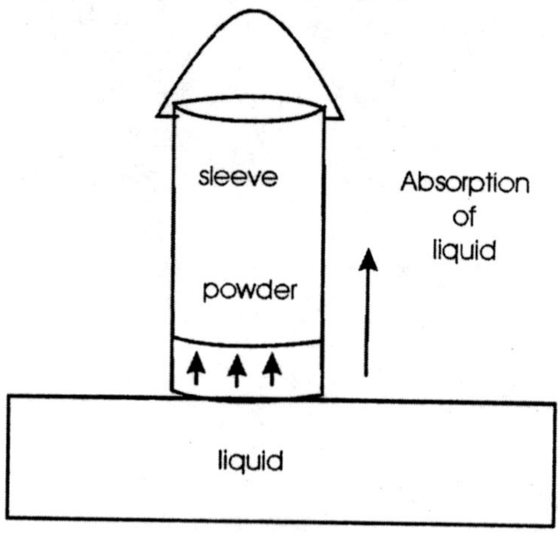


In any case the container which you use to hold your powder will also contribute to the change in forces when it contacts the surface. Especially in the first few seconds the container may contribute a significant amount to the total changes of force on the balance. It is therefore useful to be able to subtract out the contribution of the container in the adsorption graphs. Your containers may be tested, when empty, against a liquid of interest and you may then use that data file to subtract out the contributions of the empty container when testing solids against the same liquid.

**LPR 902**

Another product available from KSV is an instrument which uses much the same approach but measures the volume of the liquid absorbed instead of the mass absorbed. This approach allows for contact angles to be determined without needing to know the density of the testing liquid. This is our LPR 902 Powder Wetting Instrument. It can handle larger samples. This simple instrument allows the user to test the wetting of powders by the same Washburn method at a lower cost than the Sigma70.

## CRITICAL MICELLE CONCENTRATION

Certain molecules may be said to contain two distinct components, differing in their affinity for solutes. The part of the molecule which has an affinity for polar solutes, such as water, is said to be hydrophilic. The part of the molecule which has an affinity for non-polar solutes, such as hydrocarbons, is said to be hydrophobic. Molecules containing both types of components are said to be amphiphilic.

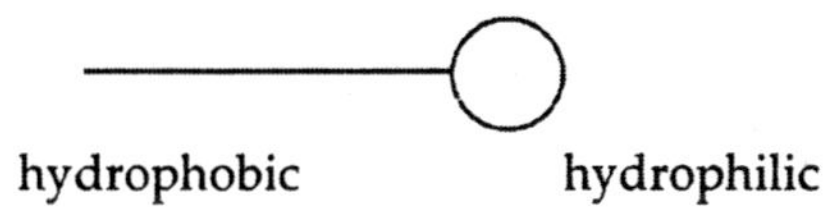


**Fig.** Amphiphilic Molecule

Such molecules display distinct behaviour when interacting with water. The polar part of the molecule 'seeks' to interact with water while the non-polar part 'shuns' interaction with water. There are two ways in which such a molecule achieve both these states.

An amphiphilic molecule can arrange itself at the surface of the water such that the polar part interacts with the water and the non-polar part is held above the surface(either in the air or in a non-polar liquid)as shown below. The presence of these molecules on the surface disrupts the cohesive energy at the surface and thus lowers the surface tension. Such molecules are called 'surface active' molecules or surfactants.

Another arrangement of these molecules can allow each component to interact with its favoured environment. Molecules can form aggregates in which the hydrophobic portions are oriented within the cluster and the hydrophilic portions are exposed to the solvent. Such aggregates are called micelles.

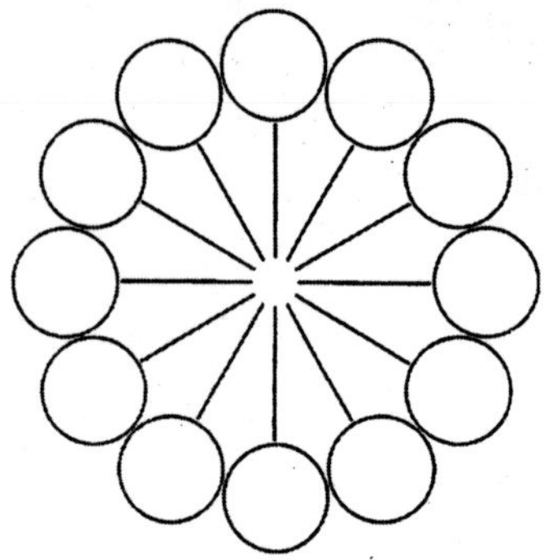

**Fig.** Spherical micelle

The proportion of molecules present at the surface or as micelles in the bulk of the liquid depends on the concentration of the amphiphile. At low concentrations surfactants will favour arrangement on the surface.

As the surface becomes crowded with surfactant more molecules will arrange into micelles. At some concentration the surface becomes completely loaded with surfactant and any further additions must arrange as micelles. This concentration is called the Critical Micelle Concentration (CMC). It follows that measurement of surface tension may be used to find CMC. A graph of surface tension vs log of concentration of surfactant added will appear as follows:

In this graph you can see three phases:

1. At very low concentrations of surfaetant only slight change in surface tension is detected.
2. Additional surfactant decreases surface tension
3. Surface becomes fully loaded, no further change in surface tension.

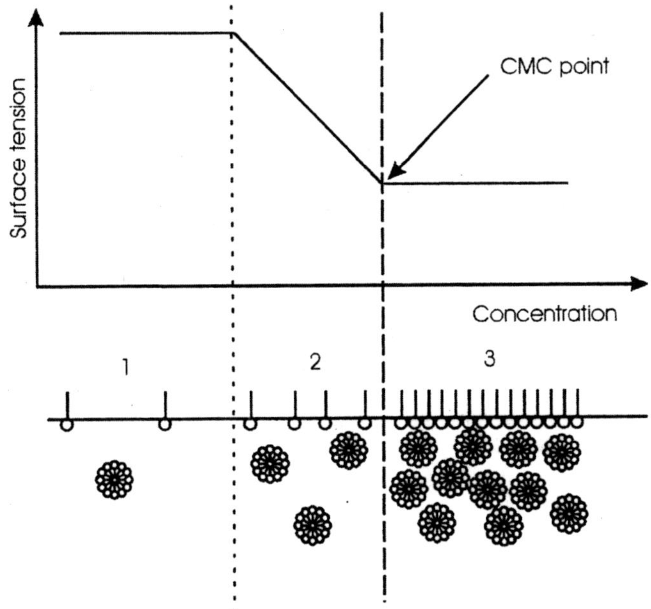


As shown above the technique for assaying CMC by measurement of surface tension is simple and straightforward. A graph of surface tension vs log concentration is produced. The CMC is found as the point at which two lines intersect; the baseline of minimal surface tension and the slope where surface tension shows linear decline.

The graph of surface tension vs log concentration may be done by measuring a series of manually mixed solutions or automatically using a KSV Sigma70 with optional dispenser. Manual mixing and testing surface tensions for a series of solutions is both time consuming and difficult. After solutions of surfactants are mixed time may be required for the surface tension to reach an equilibrium value. This time may vary from seconds to hours. When a researcher is trying to manually find CMC this may present real problems. If CMC values are found using the automated features of the Sigma 70 the stability of surface tension measurements can be easily handled. You simply enter the desired limits of stability and the maximum time allowed for measurement at any concentration and your tensiometer will continue measuring until these limits have been met. The Sigma 70 programmes also automate the mixing of concentrations for even spacing across the log scale and assist in graphics analysis to find your CMC.

A probe is chosen(Wilhelmy plate or DuNouy ring) and the measuring vessel is filled with solute. The dispenser is filled with concentrated surfactant. This short procedure is all that is required of the researcher. The Sigma 70 will automatically complete the experiment.

Your experiment will proceed in stages. First a surface tension will be measured prior to any addition. Next the first addition will be made. If you wish, you have the option of programming this as a large addition to bring the system closer to a known CMC value. The instrument will then stir your solution. Next, values of surface tension will be measured. Frequently after the addition of surfactants the values of surface tension will take some time to stabilize. You can set the values which will decide the required limits for stability. When the system has displayed the required stability in surface tension measurements or a maximum time limit has passed, your Sigma70 will make another addition and continue as above.

Additions will be made such that data will be evenly spaced along a log scale of concentration. You will select the range of concentrations assayed and the number of data points desired and the Sigma 70 will mix the appropriate solutions for measurement.

You may also select the AutoSearch option. This divides the experiment into two phases. The Search phase allows for larger additions to be made during the first portion of the experiment when data not critical to CMC calculations will occur. When a sufficient drop in ST values occurs the experiment will switch to the Measurement phase and obey the addition schedule of a normal experiment.

When all requested additions have been made the experiment will terminate. The data will be automatically stored to your hard drive and can be analysed at any later time.

**Mixing Controls:**

- Start concentration
- End concentration
- Number of concentrations assayed
- Dispensing rate
- Stirring rate
- Stirring time
- Wait after stirring
- Stability of measurements
- Maximum time measuring
- Autosearch: larger additions/ pressure drop trigger

Many researchers are interested in finding the CMC for surfactants in order to optimize detergency and minimize waste. Beyond the calculation of CMC the researcher will also produce a graph which shows surface tension across a range of concentrations. Data is also produced which shows the change in surface tension after mixing and time required to reach equilibrium.

## CHARACTERIZATION OF FIBRE WETTING

The wetting behaviour of fibres is of interest to researchers in a variety of systems. Fibre wetting applications include; the spreading of ink on paper, water repellency of fabrics, treatments on human hair, fibre reinforced composite production and many others. As the diameter of a fibre decreases the ratio of surface to volume increases and the effects of the surface on the characteristics of the fibre becomes more pronounced. In applications involving small diameter fibres the influence of the surface of the fibre may dominate most interactions.

### How is Fibre Wetting Measured?

Characterization of the wetting of fibres presents special research problems. The primary parameter reported to characterize wetting is contact angle. Typically contact angles are measured using a goniometer. This requires that a drop of the test liquid be placed on the surface of the solid tested. Placement of a drop of liquid on fibres of small diameter may be quite difficult. Strategies used to overcome this problem have included using very small drops or analyzing the shape of a drop of liquid surrounding a vertically oriented fibre. Neither of these strategies is entirely satisfactory.

A more reasonable approach for fibres is to use tensiometry to measure contact angle. The Sigma 70 tensiometer produced by KSV Instruments permits you to make rapid and reliable measurement of contact angles on fibres. This technique, commonly called the Wilhelmy method for contact angles. Briefly, a sample of your test solid is hung on the balance of the tensiometer and brought into contact with the test liquid.

When the fibre contacts the liquid the change in forces is detected and your Sigma70 registers this elevation as zero depth of immersion. As the solid is pushed into the liquid the forces on the balance are recorded. The forces on the balance are

$F_{total}$ = wetting force + weight of probe - buoyancy

Your Sigma70 has tared the weight of the probe and can remove the effects of the buoyancy force by extrapolating the graph back to zero depth of immersion. The remaining component force is the wetting force which is defined as:

Wetting force = $\gamma_{LV} P \cos\theta$

where $\gamma_{LV}$ is the liquid surface tension, P is the perimeter of the probe and $\theta$ is the contact angle. Thus at any depth data is received which can be used to calculate contact angle. This contact angle, which is obtained from data generated as the probe advances into the liquid, is called the 'advancing contact angle'. The sample is immersed to a set depth and the process is reversed. As the probe retreats from the liquid data collected is used to calculate a 'receding contact angle'. On almost all surfaces the advancing contact angle will exceed the receding contact angle. The difference between these two values, know as the contact angle hysteresis, is a subject of great interest in wetting studies and you are recommended to the references cited at the end of this note for further details.

An example of the type of graph generated in such an experiment is displayed below.

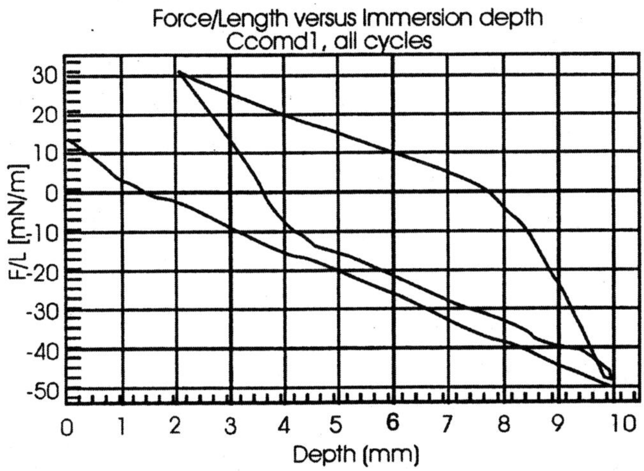


The graph displays data as wetting force vs depth of immersion. In the experiment shown two cycles of immersion/ emersion were performed. The lower slopes of the graph represent the force values as the fibre is immersed into the liquid and are used to calculate the advancing contact angle. The upper slopes of the graph represent the force values as the fibre is emersed from the liquid and are used to calculate the receding contact angle.

Mounting: KSV Instruments makes sample holders which make fibre mounting simple. If you do not have these holders the easiest way to mount small fibres may be as follows: make a small throw-away hook from light gauge wire. Attach a length of your fibre to the end of the hook using a fast drying adhesive. This will create the type of rigid anchor for your fibre which is necessary for reproducible data. Be careful not to expose any length of the fibre which will come in contact with your liquid to the adhesive. Cut off any part of the end of the fibre which has been contaminated by contact during this manipulation.

Length: The fibre sample needs to be able to penetrate the surface of your liquid and maintain a vertical orientation as it moves through the liquid. For this reason you may have to trim very small and limp fibres to less than 5mm length. Although it is useful to assay longer lengths of the sample, lengths as short as 2mm can yield good data. If your fibre is not rigid and/or wets poorly with your liquid, you will notice that it resists penetration of the surface. In such a case you should attempt to shorten the length of the sample until you achieve penetration.

## FABRICS

Analysis of the wetting of fabrics is usually treated differently than the analysis of the fibres of which they are composed. The most frequently used approach to finding contact angles for fabrics it to treat them as porous solids and use the Washburn method of analysis. An alternate approach is to study the fibres which compose the fabric. Studies have shown that the contact angle of the fabric wetting, as assayed by Washburn method equals the contact angle of the fibres of which it is composed. This appears to be true regardless of the weave of the fabric.

## ADHESION

Interest in the wetting characteristics of fibres has increased greatly in recent years. Much of this new interest involves the study of fibre reinforced composites. The successful use of fibres in composite materials depends on the adhesion of the fibres into the matrix material. The quality of the adhesion is in turn dependent on the wetting of the fibres by the matrix. Characterization of this wetting is done by contact angle experiments as described above.

## PERIMETER OF FIBRES

Fibres of a variety of sizes are easily tested for contact angles. It is not unreasonable to expect that you can get good results for fibres with diameters below 10μ. As in all cases, the accuracy of your results is directly related to the accuracy of your measurement of the perimeter of the fibre. Measurement of the perimeter of your fibre can be performed manually or by the following approach which is recomended for fibres with irregular shapes.

Perform a contact angle experiment on your fibre using a completely wetting liquid such as hexane. Remembering that:

Wetting force = $\gamma_{LV}$ P cosq

and that for completely wetting liquids $\theta = 0$, \ cosq = 1 it follows that

Wetting force = $\gamma_{LV}$ P

Using the measured wetting force and known $\gamma_{LV}$ it is simple to solve for perimeter. Please note that the wetting force used for this technique should be the receding wetting force. This force is chosen rather than the advancing force as it is more reasonably representing a situation with a zero contact angle.

## LANGMUIR AND LANGMUIR-BLODGETT FILMS

1. *Introduction:* Thin organic films of a thickness of a few nanometers (a monolayer) are the source of high expectations as being useful components in many practical and commercial applications such as sensors, detectors, displays and electronic circuit components. The possibility to synthesize organic molecules, almost without limitations, with desired structure and functionality in conjunction with a sophisticated thin film deposition technology enables the production of electrically, optically and biologically active components on a nanometer scale. An organic thin film can be deposited on a solid substrate by various techniques such as thermal evapouration, sputtering, electrodeposition, molecular beam epitaxy, adsorption from solution, Langmuir-Blodgett (LB) technique, self-assembly, etc.. The LB-technique is one of the most promising techniques for preparing such thin films as it enables

a. The precise control of the monolayer thickness,
b. Homogeneous deposition of the monolayer over large areas
c. The possibility to make multilayer structures with varying layer composition. An additional advantage of the LB technique is that monolayers can be deposited on almost any kind of solid substrate.

## SHORT HISTORY OF L- AND LB- FILMS

When describing the history of Langmuir or Langmuir-Blodgett films one is bound to start with an American statesman, Benjamin Franklin, who in 1774 reported the following to the British Royal Society:

"At length at Clapman where there is, on the common, a large pond, which I observed to be one day very rough with the wind, I fetched out a cruet of oil, and dropped a little of it on the water. I saw it spread itself with surprising swiftness upon the surface.. the oil, though not more than a teaspoonful, produced an instant calm over a space several yards square, which spread amazingly and extended itself gradually until it reached the leeside, making all that quarter of the pond, perhaps half an acre, as smooth as a looking glass."

Had Franklin made some simple quantitative calculations he would have found out that if a teaspoonful (2 ml) of oil is spread over an area of half an acre, the thickness of the film on the surface of water must be less than 2 nm.

It was not until over a hundred years later when Lord Rayleigh suspected that the maximum extension of an oil film on water represents a layer one molecule thick. At the same time the foundation for our ability to characterize monolayers on an air-water interface was set by a German woman called Agnes Pockles. She developed a rudimentary surface balance in her kitchen sink, which she used to determine (water) surface contamination as a function of area of the surface for different oils.

Irwing Langmuir was the first to perform systematic studies on floating monolayers on water in the late 1910's and early 1920's. These studies led to him being awarded the Nobel prize. As early as 1920 he reported the transfer of fatty acid molecules from water surfaces onto solid supports. However, the first detailed description of sequential monolayer transfer was given several years. These built-up monolayer assemblies are therefore referred to as Langmuir-Blodgett (LB-) films. The term "Langmuir film" is normally reserved for a floating monolayer.

After the pioneering work done by Langmuir and Blodgett it took almost half a century before scientists all around the world started to realise the opportunities of this unique technique. The first international conference on LB-films and the use of this technique has been increasing widely among scientists working on various different fields of research. Today, we are in a situation where the production of ultrathin organic films with the LB-technique has slowly started to find possible practical applications in many fields. However, even though the ideas for practical applications are growing the L and LB films are still mostly used as model systems for example for biomembrane research and multilayer coatings.

In order to understand what we are measuring and how we are measuring the properties of floating monolayers (L films) it is useful to introduce some common knowledge of surface chemistry.

## SURFACE TENSION

The molecules in a liquid have a certain degree of attraction to each other. The degree of this attraction, also called cohesion, is dependent on the properties of the substance. The interactions of a molecule in the bulk of a liquid are balanced by an equally attractive force in all directions.

The molecules on the surface of a liquid experience an imbalance of forces i.e. a molecule at the air/water interface has a larger attraction towards the liquid phase than towards the air or gas phase. Therefore, there will be a net attractive force towards the bulk and the air/water interface will spontaneously minimize its area and contract.

The net effect of this situation is the presence of free energy at the surface. The excess energy is called surface free energy and can be quantified as a measurement of energy/area. It is also possible to describe this situation as having a line tension or surface tension which is quantified as a force/length measurement. Surface tension can also be said to be a measurement of the cohesive energy present at an interface. The common units for surface tension are dynes/cm or mN/m. These units are equivalent. Solids may also have a surface free energy at their interfaces but direct measurement of its value is not possible through techniques used for liquids.

Polar liquids, such as water, have strong intermolecular interactions and thus high surface tensions. Any factor which decreases the strength of this interaction will lower surface tension. Thus an increase in the temperature of this system will lower surface tension. Any contamination, especially by surfactants, will lower surface tension. Therefore researchers should be very cautious about the issue of contamination.

## SURFACTANTS

Surface active agents (surfactants) is a larger class of molecules which have a significant technological and biological importance. Generally these molecules consist of a hydrophilic (water solubie) and a hydrophobic (water insoluble) part. This amphiphilic nature of surfactants is responsible for their association behaviour in solution (micelles, bilayers, vesicles, etc.) and their accumulation at interfaces (air/water or oil/water). The hydrophobic part usually consists of hydrocarbon or fluorocarbon chains, while the hydrophilic part consists of a polar group (-OH, -COOH, $-NH_3^+$, $-PO_4^-(CH_2)_2NH_3^+$ etc.).

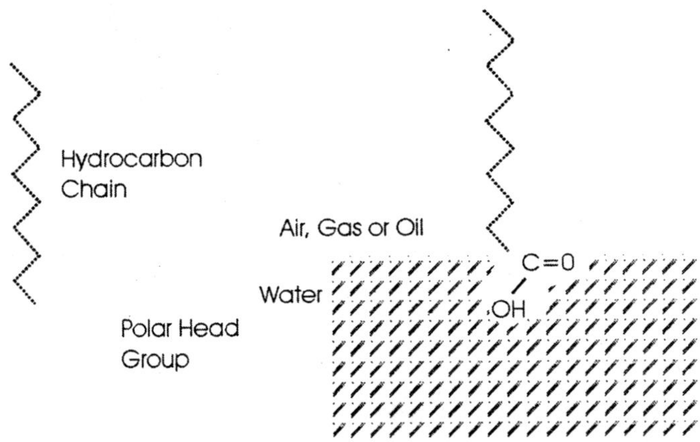


**Fig.** A schematic illustration showing the components of an amphiphile, and the orientation of an amphiphile adopted at an interfacef.

The association behaviour of surfactants in solution and their affinity for interfaces is determined by the physical and chemical properties of the hydrophobic and hydrophilic groups, respectively. The size and shape of the hydrocarbon moiety and the size, charge and hydration of the hydrophilic head group are of utmost importance in this respect. Depending on the balance between these properties a wide variety of self-assembled structures, both at interfaces and in bulk, have been observed. The driving force behind the association is the reduction of the free energy of the system. Therefore, when a surfactant comes in contact with water it accumulates at the air/water interface causing a decrease in the surface tension of water.

## INSOLUBLE MONOLAYERS

There exists a wide range of surfactants with an amphiphilic nature which drastically lower the surface tension of water. Many of these amphiphilic substances insoluble in water can with the help of a volatile and water insoluble solvent easily be spread on a water surface to form an insoluble monolayer at the air/water interface. These monolayers, also called Langmuir (L) films, represent the extreme case when considering adsorption to interfaces because all molecules are concentrated in a one molecule thick layer at the interface. The amphiphilic nature of the surfactants dictates the orientation of the molecules at the interface (air/water or oil/water) in such a way that the polar head group is immersed in the water and that the long hydrocarbon chain is pointing towards air, gas or oil.

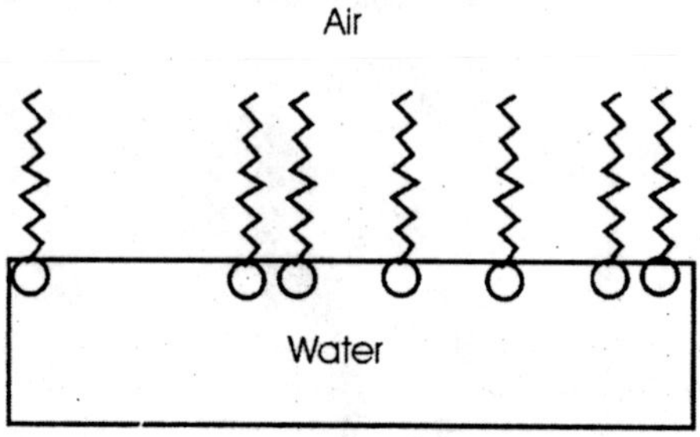


**Fig.** A schematic illustration showing a spread monolayer at the air/water interface.

The hydrocarbon chain of the substance used for monolayer studies has to be long enough in order to be able to form a insoluble monolayer. A rule of thumb is that there should be more than 12 hydrocarbons or groups in the chain ($(CH_2)_n$, $n > 12$). If the chain is shorter, though still insoluble in water, the amphiphile on the water surface tend to form micelles. These micelles are water soluble, which prevents the build-up of a monolayer at the interface. On the other hand if the length of the chain is too long the amphiphile tends to crystallize on the water surface and consequently does not form a monolayer. It is difficult to determine the optimal length for the hydrocarbon chain because its film forming ability also depends on the polar part of the amphiphile. Furthermore, the amphiphile has to be soluble in some organic solvent which is highly volatile and water insoluble (chloroform or hexane is commonly used).

## SURFACE PRESSURE

As earlier was mentioned the air/water interface possesses an excess free energy originating from the difference in environment between the surface molecules and those in the bulk. This interfacial free energy is accessible by measurements of the surface tension, $\gamma$. The surface tension of water is around 73 mN/m at 20°C, which is an exceptionally high value compared to other liquids and consequently makes water a very good subphase for monolayer studies.

When a solution of an amphiphile in a water insoluble solvent is placed on a water surface with a microsyringe, the solution spreads rapidly to cover the available area. As the solvent evapourates, a monolayer is formed. When the available area for the monolayer is large the distance between adjacent molecules is large and their interactions are weak. The monolayer can then be regarded as a two-dimensional gas. Under these conditions the monolayer has little effect on the surface tension of water. If the available surface area of the monolayer is reduced by a barrier system the molecules start to exert a repulsive effect on each other. This two-dimensional analogue of a pressure is called surface pressure, $\Pi$, and is given by the following relationship

$$\Pi = \gamma - \gamma_0$$

where $\gamma$ is the surface tension in absence of a monolayer and $\gamma_0$ the surface tension with the monolayer present.

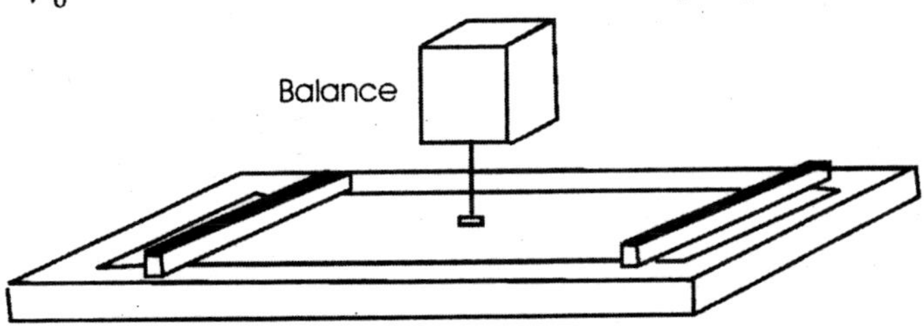


Fig. Schematic illustration of a Langmuir film balance with a Wilhelmy plate electrobalance measuring the surface pressure, and arriers for reducing the available surface area.

## THE LANGMUIR FILM BALANCE

A schematic picture of a Langmuir balance. The trough holding the subphase is usually made of Teflon® to prevent any leakage of the subphase over the edges. The trough is thermostated by circulating water in channels placed underneath the teflon trough. The surface area of the trough can be varied by sweeping movable barriers over the surface of the trough. The barrier is made of Delrin®, a hydrophilic material, and heavy enough to prevent any leakage of the monolayer beneath the barrier. The surface pressure and the mean molecular area are continuously monitored during the compression.

The surface pressure is measured by the Wilhelmy plate-method. In this method a measurement is made by determining the force due to surface tension on a plate suspended so that it is partially immersed in the subphase. This force is then converted into surface tension (mN/m or dynes/cm) with the help of the dimensions of the plate.

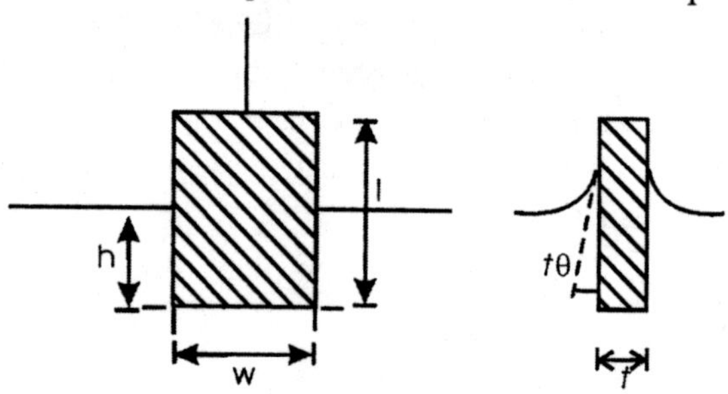


**Fig.** A Wilhelmy plate partially immersed in a water surface.

The plate is often very thin and made of platinum, but even plates made of glass, quartz, mica and filter paper can be used. The forces acting on the plate consist of the gravity and surface tension downward, and buoyancy due to displaced water upward. For a rectangular plate of dimensions $l_p$, $w_p$ and $t_p$, of material density $\rho_p$, immersed to a depth $h_l$ in a liquid of density r $_l$, the net downward force is given by the following equation:

$$F = \rho_p g l_p w_p t_p + 2\gamma (t_p w_p)(\cos\theta) - \rho_l g t_l w_l h_l,$$

where $\gamma$ is the liquid surface tension, q is the contact angle of the liquid on the solid plate and g is the gravitational constant. The surface pressure is then determined by measuring the change in F for a stationary plate between a clean surface and the same surface with a monolayer present. If the plate is completely wetted by the liquid (i.e. $\cos\theta = 1$)

the surface pressure is then obtained from the following equation:

$$\Pi = -\Delta \gamma = -[\Delta F / 2(t_p+w_p)] = -\Delta F/2w_p, \text{ if } w_p >> t_p.$$

The sensitivity can thus be increased by using a very thin plate. The force is in this way determined by measuring the changes in the mass of the plate, which is directly coupled to a sensitive electrobalance. The monolayer can also be held at a constant surface pressure, which is enabled by a computer controlled feedback system between the electrobalance and the motor responsible for the movements of the compressing barrier. This is useful when producing LB films i.e. when the monolayer is deposited on a solid substrate.

There are even other ways to control the area of the monolayer and to measure the surface pressure but the constructions above are the most commonly used.

## SURFACE PRESSURE - AREA ISOTHERMS

The most important indicator of the monolayer properties of an amphiphilic material is given by measuring the surface pressure as a function of the area of water surface available to each molecule. This is carried out at constant temperature and is known as a surface pressure - area isotherm or simply "isotherm". Usually an isotherm is recorded by compressing the film (reducing the area with the barriers) at a constant rate while continuously monitoring the surface pressure. Depending on the material being studied, repeated compressions and expansions may be necessary to achieve a reproducible trace.

A number of distinct regions is immediately apparent on examining the isotherm. These regions are called phases. As one can see when the monolayer is compressed it can pass through several different phases which are identified as discontinuities in the isotherm. The phase behaviour of the monolayer is mainly determined by the physical and chemical properties of the amphiphile, the subphase temperature and the subphase composition. For example, various monolayer states exist depending on the length of the hydrocarbon chain length and the magnitude of other cohesive and repulsive forces existing between head groups. An increase in the chain length increases the attraction between molecules, condensing the π-A-isotherm. On the other hand, if an ionasible amphiphile is used the ionisation of the head groups induces repulsive forces tending to oppose phase transitions.

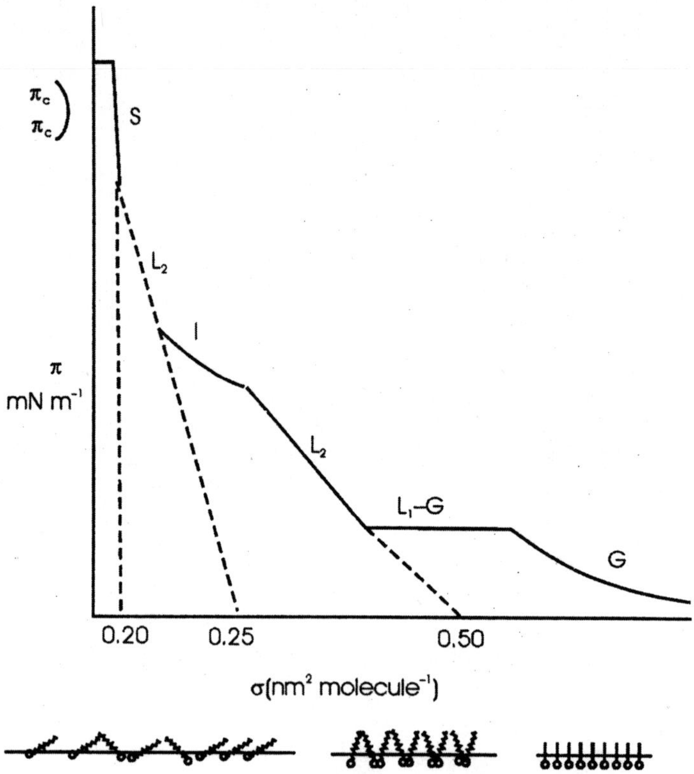


**Fig.** Schematic p-A-isotherm() and B) Orientation of the molecules in different phases().

A simple terminology used to classify different monolayer phases of fatty acids. At large the monolayers exist in the gaseous state (G) and can on compression undergo a phase transition to the liquid-expanded state ($L_1$). Upon further compression, the $L_1$ phase undergoes a transition to the liquid-condensed state ($L_2$), and at even higher densities the monolayer finally reaches the solid state (S). If the monolayer is further compressed after reaching the S state the monolayer will collapse into three-dimensional structures. The collaps is generally seen as a rapid decrease in the surface pressure or as a horisontal break in the isotherm if the monolayer is in a liquid state.

There are also many other critical points in a π-A-isotherm such as the molecular area at which an initial, pronounced increase in the surface pressure is observed, $A_i$, and the surface pressures at which phase transitons occure between the $L_1$ and $L_2$ state and the $L_2$ and S state.

Typical isotherms of a fatty acid with a single hydrocarbon chain and a phospholipid with two hydrocarbon chains (right). Following the definitions above one can see that the fatty acid has three distinct regions gas (G), liquid ($L_1$) and solid (S), while the phospholipid has an additional almost horizontal transition phase ($L_2$-$L_1$) between the two different liquid phases. This is very common for phospholipids and the position of this horizontal transition phase is very temperature dependent. As the temperature is increased the surface pressure value at which the horizontal transition phase occurs will increase and vice versa.

## DEPOSITION - LB FILMS

Apart from being used as a tool for monolayer studies the Langmuir film balance can also be used for building up highly organised multilayers of the used amphiphile. This is accomplished by successively dipping a solid substrate up and down through the monolayer while simultaneously keeping the surface pressure constant by a computer controlled feedback system between the electrobalance measuring the surface pressure and the barrier moving mechanism. Consequently the floating monolayer is adsorbed to the solid substrate. In this way multilayer structures of hundreds of layers can be produced. These multilayer structures are commonly called Langmuir-Blodgett or simply LB films.

The LB deposition is traditionally carried out in the "solid" phase. The surface pressure is then high enough to ensure sufficient cohesion in the monolayer, e.g. the attraction between the molecules in the monolayer is high enough so that the monolayer does not fall apart during transfer to the solid substrate. This also ensures the build up of homogeneous multilayers. The surface pressure value that gives the best results depends on the nature of the monolayer and is usually established empirically. However, amphiphiles can seldom be successfully deposited at surface pressures lower than 10 mN/m, and at surface pressures above 40 mN/m collapse and film rigidity often pose problems. When the solid substrate is hydrophilic (glass, $SiO_2$ etc.) the first layer is deposited by raising the solid substrate from the subphase through the monolayer, whereas if the solid substrate is hydrophobic (HOPG, silanized $SiO_2$ etc.) the first layer is deposited by lowering the substrate into the subphase through the monolayer.

There are several parameters that affect on what type of LB film is produced. These are the nature of the spread film, the subphase composition and temperature, the surface pressure during the deposition and the deposition speed, the type and nature of the solid substrate and the time the solid substrate is stored in air or in the subphase between the deposition cycles.

The quantity and the quality of the deposited monolayer on a solid support is measured by a so called transfer ratio. This is defined as the ration between the decrease in monolayer area during a deposition stroke, $A_l$, and the area of the substrate, $A_s$. For ideal transfer the t.r. is equal to 1. Depending on the behaviour of the molecule the solid substrate can be dipped through the film until the desired thickness of the film is achieved. Different kind of LB multilayers can be produced and/or obtained by successive deposition of monolayers on the same substrate. The most common one is the Y-type multilayer, which is produced when the monolayer deposits to the solid substrate in both up and down directions. When the monolayer deposits only in the up or down direction the multilayer structure is called either Z-type or X-type.

Intermediate structures are sometimes observed for some LB multilayers and they are often referred to be XY-type multilayers.

The production of so called alternating layers which consist of two different kind of amphiphiles is also possible by using highly sophisticated instruments. In such an instrument there is a trough with two separate compartments both possessing a floating monolayer of a different amphiphile. These monolayers can then be alternatingly deposited on one solid substrate.

An alternative way to deposit the monolayer is the Langmuir-Schaeffes (LS) technique. This technique differs from the vertical technique described above only in the sense that the solid substrate is horizontally lowered in contact with the monolayer.

## APPLICATIONS OF SURFACE CHEMISTRY

Environmental surface chemistry involves processes that occur at the interface between the regolith, hydrosphere and atmosphere. The more limited scope of the present review addresses natural and anthropogenically-induced inorganic geochemical reactions between solutes in surface and ground waters and soil and aquifer substrates. Important surficial reactions include sorption, ion exchange, dissolution, precipitation and heterogeneous oxidation/reduction processes occurring at the solid/aqueous interface. Recent research advances in this field have addressed, both directly and indirectly, societal issues related to water quality, pollution, biogeochemical cycling, nutrient budgets and chemical weathering related to long term global climate change. This review will include recent advances in the fundamental and theoretical understanding of these surficial processes, breakthroughs in experimental and instrumental surface characterization, and development of methodologies for field applications.

## REACTION MECHANISMS

In the last several years, significant advancements have been made in understanding fundamental mechanisms controlling sorption, precipitation, dissolution, and heterogeneous redox reactions. This work points to a greater unification of our understanding of these surface processes by emphasizing similarities based on quantum mechanical, thermodynamic and kinetic approaches. Atomistic theory may ultimately explain the fundamental bonding and atomic dynamics of minerals and their interaction with hydrated ions and waters. Performed *ab-initio* calculations of interatomic forces and activation energies for quartz that were in accord with independently-derived experimental results. At present the structures of most geochemical interfaces and the nature of the activated complex are too complicated for such treatments. Recent advances include the incorporation of crystallochemical models that describe surface reaction sites not only in terms of the number and nature of metal ions bonded to them but also the nature of the hydrogen bonds that the surface can donate.

Efforts to quantify observed sorption in both natural and perturbed geochemical systems have increasingly relied on numerical surface complexation models. In an extensive review of the subject, point out that the basic tenets of surface complexation theory are that;

1. The surface is composed of specific functional groups that react with dissolved solutes to form surface complexes,
2. Equilibrium can be described via mass action laws incorporating electrostatic conditions,
3. Surface charge is treated as a consequence of distribution of the functional groups
4. The empirically derived equilibrium constants are related to thermodynamic activities of the surface species. Recently surface complexation models have also been expanded to include the description of ion solvation that permits more accurate description of specific interaction with complex oxides and silicates. Most direct applications to environmental issues have been concerned with defining intrinsic parameters

needed to fit experimental data to surface complexation models.

Significant advances have also defined the transition between sorption processes and the subsequent development of solid solutions in which the sorbate becomes incorporated into the substrate structure. Such distinctions are important in explaining the apparent irreversibility of many sorption processes and the thermodynamic solubility controls on aqueous systems. Several workers have investigated metal sorption/solid solution transitions in Fe and Mn oxides, important substrates in the environment. In the first study, electron spin resonance (ESR) was used to document a progressive transition from initial weak interaction to solid solution formation as a function of increasing $Mn^{2+}$ on the surface of $FeCO_3$. In the second study, surface sensitive techniques indicated initial surface $Cd^{2+}$ sorption that, with time, resulted in solid state diffusion and solid solution formation within the calcite substrate..

Although surface complexation was originally developed to explain sorption processes, the same theory is applicable to describing the effects of pH and organic and inorganic ligands on the dissolution of oxides and silicates. The dissolution of albite to be directly proportional to the surface concentration of positively charged aluminum sites indicating kinetics controlled by the rate of hydrolysis of Al-O-Si bridging bonds. Using complexation modeling, that dissolution of kaolinite occurred from both edge sites and the gibbsite surface and that the relative rates varied differently as a function of pH. The role of protons, carbonic acid and ligand complexation on the dissolution rates of calcite.

When the surface free energy is zero, rates of both dissolution and precipitation reactions are equal, which is equivalent to macroscopic equilibrium. Based on transition state theory, microscopic reversibility or detailed balancing at small deviations from equilibrium should result in a linear relationship between the deviation in free energy and the dissolution or precipitation rate. Recent work investigating dissolution of gibbsite, kaolinite, and albite have generally validated this theory that close to equilibrium, a direct proportionality exists between free energy and dissolution rate. Far from equilibrium, rates were found to be independent of free energy. The most impressive feature of these experiments is the dramatic increase in the rate as a certain critical undersaturation is reached. This change in rate law is linked to the density of etch pits and surface defects, the formation of which have been predicted to form based on Monte Carlo simulations at critical unsaturations. Experimental work on a variety of minerals with varying dislocation densities has supported this theoretical prediction. Most silicate dissolution experiments have been conducted under highly undersaturated conditions. However natural weathering may occur much closer to equilibrium, thus explaining commonly observed slower dissolution rates for natural systems.

Significant advances have been made in the field of heterogeneous redox reactions involving the transfer of electrons between solid and aqueous phases. Such processes are environmentally important because the oxidation state of many aqueous species, such as transition metals, determine their solubility and sorption characteristics. Research in this field can be categorized as to whether oxidation or reduction occurs at surface substrate. The oxidation of Fe(II), V(IV) Mn(II), and Cu(I) by dissolved oxygen has been shown to be favoured thermodynamically and kinetically by specific adsorption to hydrous ferric oxide and Al surfaces. Hydroxide ligands increase electron densities near the reduced metal ion and stabilize the increased oxidation state. Several studies have recently addressed the role of heterogeneous electron transfer related to reductive dissolution and precipitation of oxides both in the presence of inorganic and organic reducing agents.

Surfaces of mineral phases that contain reduced species such as Fe(II) can also serve as electron donors during redox reactions. White presented an extensive review of heterogeneous reduction of aqueous species such as Cr(VI), Cu(II), V(V), and Np (VI) by Fe(II) minerals and presented a detailed characterization of Fe(II) oxide electrochemical reactions. Important from an environmental prespective is the role of Fe(II) oxides and sulfides in the reductive reaction of aqueous organic compounds as demonstrated for carbon tetrachloride degradation at the surfaces of pyrite and mica. Significant progress has also continued on understanding photo-reduction reaction on mineral surfaces. Also photocatalytic oxidation of organics such as chlorophenol has been demonstrated to occur on substrate surfaces such as $TiO^2$.

## ANALYTICAL ADVANCES

A necessary complement to macroscopic studies dealing with sorption, redox, precipitation and dissolution reactions are microscopic measurements capable of providing direct, molecular-level information on the composition and structure of the surface substrate. Many of the recent major advances in understanding the surface chemistry of geochemical systems have been driven by application of surface-sensitive analytical techniques. Application continued on the use of ultra-high vacuum spectroscopic techniques including the use of Auger electron spectroscoppy (AES) and X-ray photoelectron spectroscopy (XPS) for surface characterization. Examples include the characterization of surface chemistry of weathering basalt flows, the use of XPS and secondary ion mass spectroscopy (SIMS) to characterize feldspar surfaces and the use of XPS to study sorption/desorption reactions of Hg(II) on Fe sulfide surfaces. Applications of nuclear magnetic resonance (NMR) include 29Si studies on the characterization of coordination bonding of water on $SiO_2$ surfaces, and the use of Cs to study bonding sites of 133Cs on clays. Also significant advances have been made in high resolution transmission electron microscopy (HRTEM) to characterize submicroscopic isovolumetric reactions associated with the transformation of primary silicates to clay minerals.

The first paper describing the application of scanning tunneling microscopy (STM) to the semiconducting minerals hematite and galena, an increasing number of studies have documented atomic and molecular resolution on geochemical substrates using this technique and the allied technique atomic force microscopy (AFM). Both techniques make use of recent advances in piezoelectric translators that can bring sharp metallic or crystalline tips within several Angstroms of the surface. For STM, the changes in the tunneling current established by a bias voltage between the tip and surface permit mapping of the density and energy of the electron states and topography of conducting and semiconducting minerals. For AFM, the same piezoelectric platform is used but Van der Waals, electrostatic or magnetic forces between the tip and sample are sensed through deflection of a cantilever. Recent specialized implementations of AFM include lateral or frictional force microscopy and magnetic force microscopy. These technologies are currently just being recognized in geochemical applications. The resolution of the STM technique has been demonstrated by a series of papers detailing step formation, oxidation, and sorption on sulfide and Fe oxide minerals with the later reference documenting real time adsorption-desorption coupled with outer-sphere diffusion of Cr(III) on -$\alpha Fe_2O_3$.

In AFM analysis, interaction between the tip and sample is more complex than in STM analysis and therefore true atomic resolution is not readily obtained. The principal geochemical applications to date have centreed on detailed 3-dimensional topographic mapping of mineral surfaces and resolution of structural features including experimental validation of ledge motion models for the dissolution of quartz and in situ observation of the dynamics of monomolecular growth steps on calcite.

The increased availability of synchrotron radiation sources has resulted in a significant increase in application of a number of techniques using X-ray absorption spectroscopy (XAS). These methods have distinct advantages over more established approaches such as XPS and AES in that XAS techniques do not require a vacuum. Techniques applied to geologic materials include extended X-ray absorption fine line (EXAFS) and X-ray absorption near edge structure (XANES). A detailed of these techniques applied to geologic substrates. Experiments consist of exposing a sample to a monochromatic beam of X-rays that is scanned through a range of energies from below to above the absorption edge of an element. The technique is sensitive to the electronic structure of two or three closest shells of atomic neighbours surrounding the absorbing atom and thus provides information on the composition of solid under in situ conditions in contact with water. The method is a bulk analytical technique in which the structural information is summed over all environments of a given absorber. However characterization of surface species and complexation is commonly done on samples with high surface areas and sorbent concentrations and low internal concentrations.

Recent studies detailing sorption processes using these techniques include the application of XANES and EXAFS to the $\lambda$-$Al_2O_3$/water interface. Work of Chisholm-Brause demonstrated that Pb(II) bonds directly to the surface as inter-sphere complexes and that the number and size of these complexes increases with surface coverage. EXAFS studies of $\alpha$ FeOOH likewise showed that Np (V) bonds as inter-sphere complexes with a lack of any ordered Np oxide or hydroxide structure denoting coprecipitation. Studies of the surface chemistry of ferrihydrite using EXAFS showed clear evidence for inner-sphere bidentate sorption of arsenate. With increasing As/Fe ratios, the ferrihydrite structure became disordered denoting surface poisoning and a decrease in the extent of crystallization.

## APPLICATIONS TO ENVIRONMENTAL SYSTEMS

The ultimate objective of mechanistic and analytical surface studies is to better characterize geochemical processes in the natural environment. Such application must also account for complexities not encountered in controlled experimental studies including spatial and chemical heterogeneity and time spans that can often not be simulated under laboratory conditions. Such complexities most often require that field studies involve significant and long-term financial and personnel commitments. These obstacles are principally responsible for the limited number of studies that have detailed surface geochemical processes under field conditions. Such information is however extremely important in validating experimental results and for calibrating hydrochemical transport models.

Studies on the effects of pH perturbations on sorption and dissolution of metals in stream waters have been conducted by McKnight and Bencala and Fuller and Davis. In the first study, artificial acidification of a stream resulted in the dissolution of heterogeneous hydrous ferric oxides that where then reduced by solution phase photochemical reactions. In the latter study, natural pH perturbations, due to natural photosynthetic processes, were shown to control arsenate sorption on iron hydroxides and diurnal cycling in stream water. Extensive artificial tracer experiments in a shallow contaminated aquifer at Cape Cod demonstrated that chromium (VI) transport was controlled by reduction to Cr(III) and irreversible sorption on the sand and gravel substrate. These studies demonstrate the importance of hydrologic and redox heterogeneity in controlling the spatial distributions of the tracer plumes. Heterogeneous redox processes were also shown to be the controlling factors in selenium contamination in groundwater. In the case of the Kesterson Reservoir in California, plumes of oxidizing surface water were shown to counteract the natural reducing groundwater environment permitting local mobility of Se(VI). Elsewhere in California, Dubrovsky found steady groundwater redox conditions, defined by a change in aquifer mineralogy, was the principal control of Se(VI) mobility. In both cases, the selenium was removed from the aqueous phase because of reduction to elemental selenium on solid substrates. All of the above examples demonstrate the importance of detailed geochemical and hydrologic characterization in order to define important surficial chemcial processes. Such information is critical in making informed and effective descisions concerning significant enviromental issues facing society.

# 6

# High Surface Area Materials

## INTRODUCTION

The trend to smaller and smaller structures, that is, miniaturization, is well known in the manufacturing and microelectronics industries, as evidenced by the rapid increase in computing power through reduction on chips of the area and volume needed per transistor. In the materials area this same trend towards miniaturization also is occurring, but for different reasons. Smallness in itself is not the goal. Instead, it is the realization, or now possibly even the expectation, that new properties intrinsic to novel structures will enable breakthroughs in a multitude of technologically important areas.

Of particular interest to materials scientists is the fact that nanostructures have higher surface areas than do conventional materials. The impact of nanostructure on the properties of high surface area materials is an area of increasing importance to understanding, creating, and improving materials for diverse applications. High surface areas can be attained either by fabricating small particles or clusters where the surface-to-volume ratio of each particle is high, or by creating materials where the void surface area (pores) is high compared to the amount of bulk support material. Materials such as highly dispersed supported metal catalysts and gas phase clusters fall into the former category, and microporous (nanometer-pored) materials such as zeolites, high surface area inorganic oxides, porous carbons, and amorphous silicas fall into the latter category.

There are many areas of current academic and industrial activity where the use of the nanostructure approach to high surface area materials may have significant impact:

- Microporous materials for energy storage and separations technologies, including nanostructured materials for highly selective adsorption/separation processes such as $H_2O$, $H_2S$, or $CO_2$ removal; high capacity, low volume gas storage of $H_2$ and $CH_4$ for fuel cell applications and high selectivity; high permeance gas separations such as $O_2$ enrichment; and $H_2$ separation and recovery
- Thermal barrier materials for use in high temperature engines
- Understanding certain atmospheric reactions
- Incorporation into construction industry materials for improved strength or for fault diagnostics
- Battery or capacitor elements for new or improved operation
- Biochemical and pharmaceutical separations
- Product-specific catalysts for almost every petrochemical process

In catalysis the key goal is to promote reactions that have high selectivity with high yield. It is anticipated that this goal will be more closely approached through tailoring a catalyst particle via nanoparticle synthesis and assembly so that it performs only specific chemical conversions, performs these at high yield, and does so with greater energy efficiency. In the electronics area one may anticipate manufacture of single electron devices on a grand scale. Manufacture of materials with greatly improved properties in one or more areas such as strength, toughness, or ductility may become commonplace. In separations science new materials with well defined pore sizes and high surface areas are already being fabricated and tested in the laboratory for potential use in energy storage and separations technologies. In addition, many laboratories around the world are actively pursuing the potential to create novel thermal barrier materials, highly selective sensors, and novel construction materials whose bonding and strength depend upon the surface area and morphology of the nanoscale constituents. Many are also engaged in developing molecular replication technologies for rapid scaleup and manufacturing.

The nanoscale revolution in high surface area materials comes about for several reasons. The late 1970s the scientific community has experienced enormous progress in the synthesis, characterization, and basic theoretical and experimental understanding of materials with nanoscale dimensions, i.e., small particles and clusters and their very high surface-to- volume ratios. Second, the properties of such materials have opened a third dimension to the periodic table, that is, the number of atoms, N. N now becomes a critical parameter by which the properties for "small" systems are defined. As a simple example, for metals we have known for decades that the atomic ionization potential (IP) is typically about twice the value of the bulk work function. It is only relatively recently that experiments have shown that the IP (and electron affinity) for clusters containing a specific number N of (metal) atoms varies dramatically and non-monotonically with N for clusters containing less than 100-200 atoms. Other properties such as chemical reactivity, magnetic moment, polarizability, and geometric structure, where they have been investigated, are also found to exhibit a strong dependence on N. Expectations for new materials with properties different from the atom or the bulk material have been realised. The opportunity is now open to precisely tailor new materials through atom-by-atom control of the composition (controlling the types as well as the numbers of atoms) in order to generate the clusters or particles of precision design for use in their own right or as building blocks of larger-scale materials or devices-that is, nanotechnology and fabrication at its ultimate.

Such precision engineering or tailoring of materials is the goal of much of the effort driving nanoscale technology. Scientists and engineers typically have approached the synthesis and fabrication of high surface area nanostructures from one of two directions:

1. The "bottom up" approach in which the nanostructures are built up from individual atoms or molecules. This is the basis of most "cluster science" as well as crystal materials synthesis, usually via chemical means. Both high surface area particles and micro- and mesoporous crystalline materials with high void volume (pore volume) are included in this "bottom up" approach.
2. The "top down" approach in which nanostructures are generated from breaking up bulk materials. This is the basis for techniques such as mechanical milling, lithography, precision engineering, and similar techniques that are commonly used to fabricate nanoscale materials, which in turn can be used directly or as building blocks for macroscopic structures.

A fundamental driving force towards efforts to exploit the nanoscale or nanostructure is based upon two concepts or realizations:

1. That the macroscopic bulk behaviour with which we are most familiar is significantly different from quantum behaviour, and
2. That materials with some aspect of quantum behaviour can now be synthesized and studied in the laboratory.

Obviously, quantum behaviour becomes increasingly important as the controlling parameter gets smaller and smaller. There are numerous examples of quantum behaviour showing up in high surface materials: the fact introduced above that clusters are found to exhibit novel (compared to the bulk) electronic, magnetic, chemical, and structural properties; the fact that the diffusivity of molecules through molecular sieving materials cannot be predicted or modeled by hard sphere molecule properties or fixed wall apertures;

and the fact that catalysts with one, two, or three spatial dimensions in the nanometer size range exhibit unique catalytic or chemical activity.

## OPPORTUNITIES FOR CLUSTERS AND NANOCRYSTALLINE MATERIALS

Clusters are groups of atoms or molecules that display properties different from both the smaller atoms or molecules and the larger bulk materials. Many techniques have been developed to produce clusters, beams of clusters, and clusters in a bottle for use in many different applications including thin film manufacture for advanced electronic or optical devices, production of nanoporous structures as thermal barrier coatings, and fabrication of thin membranes of nanoporous materials for filtration and separation.

The unique properties of nanoparticles make them of interest. For example, nanocrystalline materials composed of crystallites in the 1-10 nm size range possess very high surface to volume ratios because of the fine grain size. These materials are characterized by a very high number of low coordination number atoms at edge and corner sites which can provide a large number of catalytically active sites. Such materials exhibit chemical and physical properties characteristic of neither the isolated atoms nor of the bulk material. One of the key issues in applying such materials to industrial problems involves discovery of techniques to stabilize these small crystallites in the shape and size desired. This is an area of active fundamental research, and if successful on industrially interesting scales, is expected to lead to materials with novel properties, specific to the size or number of atoms in the crystallite.

A typical objective of nanoscale catalyst research is to produce a material with exceedingly high selectivity at high yield in the reaction product or product state, i.e., chemicals by design, with the option of altering the product or product state simply by changing the surface functionality, elemental composition, or number of atoms in the catalyst particle. For instance, new catalysts with increasing specificity are now being fabricated in which the stoichiometry may be altered due to size restrictions or in which only one or two spatial dimensions are of nanometer size.

Five recent examples where nanocrystalline metallic and ceramic materials have been successfully investigated for catalytic applications.

## NOVEL CATALYTIC PROPERTIES OF NANOSTRUCTURED GOLD CATALYSTS

In the study of transition metal catalytic reactions the group at Osaka National Research Institute has discovered that nanoscale gold particles display novel catalytic properties. Highly selective low temperature catalytic activity is observed to switch on for gold particles smaller than about 3-5 nanometers in diameter. Accompanying this turn on in catalytic activity is the discovery that these nanoscale gold particles (crystals) also have icosahedral structure and not the bulk fcc structure, again a nanoscale phenomena not available with bulk samples.

In fabricating these novel catalytic materials several issues appear to be crucial. For instance, the Osaka group has shown that (a) the preparation method plays a crucial role for generating materials with high catalytic activity and selectivity; that (b) the catalytic activity, selectivity and temperature of operation is critically dependent on the choice of catalyst support, and that (c) water (moisture) even in parts per million (ppm) levels dramatically alters the catalytic properties. Figure shows the effect of moisture on the conversion profiles for CO oxidation for nanoscale gold catalysts supported on cobalt oxide.

Examples of novel catalytic behaviour of nanoscale gold particles include the following:

- CO oxidation at temperatures as low as -70°C, as illustrated in Figure
- Very high selectivity in partial oxidation reactions, such as 100% selectivity at 50°C for oxidation of propylene to propylene oxide as well as near room

temperature reduction of nitric oxide with $H_2$ using alumina-supported gold nanoparticles

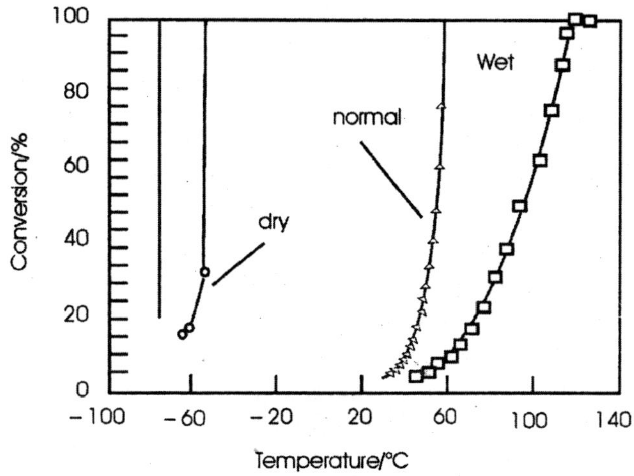


**Fig.** Effect of moisture on conversion profiles for CO oxidation over $Co_3O_4$ and $Au/Co_3O_4$.

## THE INDUSTRIALLY IMPORTANT HDS REACTION

For catalysts based on the layered compound $MoS_2$, maximum hydrodesulfurization (HDS) activity is obtained only on well-crystallized nanosized materials, while HDS selectivity is determined by the number of layers or "stack height" of the nanocrystalline $MoS_2$. In the hydrodesulfurization reaction, cyclohexylbenzene occurs only on the $MoS_2$ "rim sites," or those around the edges of the stack, whereas the pathway to biphenyl requires both "rim" and "edge" sites. Thus, the reaction selectivity can be controlled by controlling the aspect ratio of the nanoparticles of $MoS_2$. Such control of one- and two-dimensional nanostructures for selective chemical advantage is an exciting new area of research. Of course, a major industrial challenge will be to fabricate such nanocrystals in a commercializable form.

The cerium oxide ($CeO_{2-x}$) materials have been found to possess a significant concentration of $Ce^{3+}$ and oxygen vacancies, even after high temperature (500°C) calcination.

Such nanoclusters give rise to a substantial reduction in the temperature of selective $SO_2$ reduction by CO and exhibit excellent poisoning resistance against $H_2O$ and $CO_2$ in the feed stream compared to that for conventional high surface area cerium oxide.

## ELECTROCHEMICAL REDUCTION OF METAL SALTS

Electrochemical reduction of metal salts is yet another option available to control the size of nanoscale catalyst particles. This has been successfully used to prepare highly dispersed metal colloids and fix the metal clusters to the substrate. Control of the current density during the electrochemical synthesis process allows one to control the size

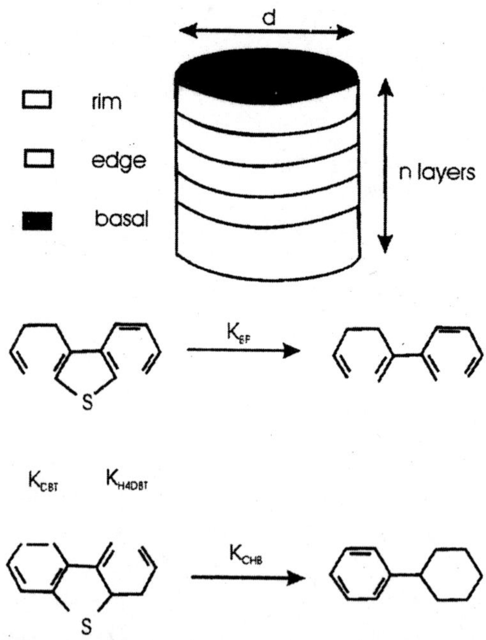


**Fig.** Hydrodesulfurization reaction. Selective catalysis is controlled by either the edge or rim of $MoS_2$.

of nanoscale transition metal particles. A combination of scanning tunneling microscopy (STM) and high-resolution transmission electron microscopy (TEM) has allowed surfactant molecules to be visualized on nanostructured palladium clusters.

## ENERGY STORAGE ENHANCED BY CATALYTIC DISSOCIATION OF $H_2$

Materials with higher hydrogen storage per unit volume and weight are considered by many to be an enabling technology for vehicular fuel cell applications. Scientists at Los Alamos National Laboratories have developed an approach that enables materials such as Mg to be used for hydrogen storage. Magnesium is of interest because it can store about 7.7 wt % hydrogen, but its adsorption/desorption kinetics are slow, i.e., the rate of charge (hydrogen dissociation and hydride formation) is much slower than in metal hydrides. At Los Alamos, high surface area mixtures of nanoscale Mg and $Mg_2Ni$ particles are produced by mechanical means, ball milling. The addition of $Mg_2Ni$ catalyzes the $H_2$ dissociation such that the rate of hydrogen adsorption increases to a rate comparable to that of $LaNi_5$. Once a hydride is formed, the hydrogen "spillover" leads to magnesium hydride formation.

Figure illustrates the hydrogen adsorption/desorption characteristics of the mixture of Mg with 23 atomic % Ni. As can be seen, a low pressure plateau at about 1500 torr is obtained for this particular sample. Experiments show that the pressure plateau can be tailored through such alloying. Studies with other nanoscale materials, including other catalysts such as FeTi and $LaNi_5$, are presently ongoing to further improve both the capacity and the rate of hydrogen storage.

In self-assembly large molecular structures are obtained from the organization of a large number of molecules or atoms into a given shape, typically through specific interactions of the molecules among themselves and with a template. The interaction of the different bonding mechanisms is an area of strong fundamental research interest. Only two areas will be highlighted here: zeolites and carbon materials. Both of these materials exhibit characteristics of self-assembly, namely novel and reproducible structures that can be fabricated in industrially significant quantities.

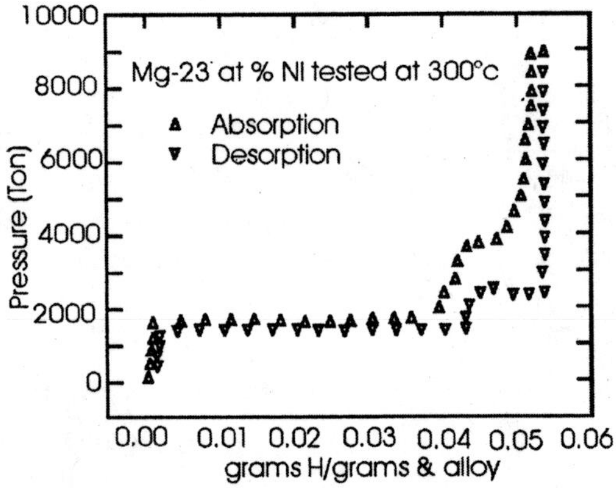


**Fig.** Hydrogen absorption-desorption characteristics for mixture of Mg and $Mg_2Ni$ prepared by mechanical alloying.

## ZEOLITIC MATERIALS

Aluminosilicates (e.g., zeolites) are crystalline porous nanostructures with long range crystalline order with pore sizes which can be varied from about 4 Å to 15 Å in conventional zeolites. Figure shows a 3-dimensional (MFI) zeolite cage structure together with a depiction of the straight and ziz-zag channels and a 2-dimensional zeolite with channels only in 2 directions. The vertices in the stick drawing denote position of the O atoms in the crystalline lattice. This particular zeolite has 10 atoms in the zeolite "window." The size of the window is determined by the number of oxygens in the ring. Table gives approximate window dimensions for zeolites as a function of the number of oxygens in the ring.

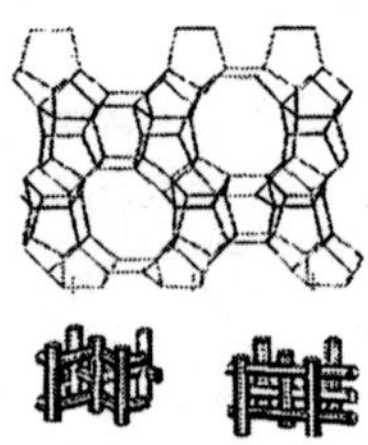

**Fig.** Typical zeolite structures together depicting the positions of the O atoms (vertices in upper figure) and two different zeolitic structures one with a three dimensional structure and a zeolite with a two dimensional channel structure.

**Table.** Zeolite Channel "Window" Dimension for Number of Oxygens in Ring

| Number of Oxygens in Ring | Ring Diameter (Å) |
|---|---|
| 4 | 1.2 |
| 5 | 2.0 |
| 6 | 2.8 |
| 8 | 4.5 |
| 10 | 6.3 |
| 12 | 8.0 |

As can be seen by examination of Table, molecules can pass through or be blocked from transport through or into the zeolite depending on the zeolite. For example normal hexane with a kinetic molecular diameter of about 5.1 Å can pass through a 10 ring or larger, whereas cyclohexane with a kinetic molecular diameter of 6.9 Å would be hard pressed to pass through a 10 ring. Thus all other things being equal, a 10-ring zeolite could be used to separate mixtures of normal hexane and cyclohexane. It is this property together with the ability to chemically modify the acidity of zeolitic materials that makes them extremely valuable as selective sorbants, as membranes and for use in selective catalytic reactions.

A new family of aluminosilicates (M41S) with pores sizes between 20 and 100 Å in diameter were reported by Mobil researchers. One of particular interest is MCM-41, which consists of hexagonal arrays of uniform 2 to 10 nanometer-sized cylindrical pores. Not only can such materials be synthesized, but novel structures such as "tubules-within-a-tubule" have been fabricated as mesoporous molecular sieves in MCM-41. Of particular interest is the possibility of expanding the so-called "liquid crystal templating" mechanism to non-aluminum dopants within the silicate MCM-41 framework and to derive non-siliceous MCM-41 type of materials.

## CARBON MATERIALS

The carbon-based materials of interest from a molecular self-assembly point of view include fullerenes and their relatives, including endohedral fullerenes and metal-coated fullerenes, carbon nanotubes, carbon nanoparticles, and porous carbons. The discovery of techniques to produce soluble carbon in a bottle, research on and with carbon materials has skyrocketed. Not only can the molecular forms of carbon be synthesized, characterized, and studied for applications, but many other new carbon materials such as multi- and single-walled carbon nanotubes can now be produced in macroscopic quantities. Figure illustrates the broad variety of carbon nanotube structures whose properties are now being examined both theoretically and experimentally. A rich literature on these new carbon materials now exists. This report will only attempt to highlight a few important recent examples in the area of high surface area materials.

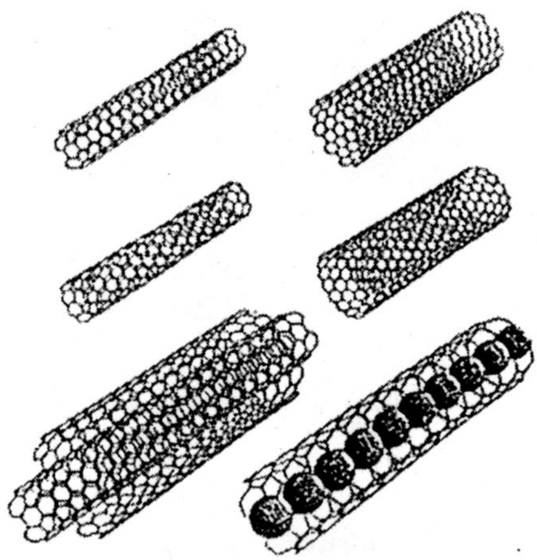

**Fig.** Examples of carbon nanotube structures, including multiwalled and metal-atom-filled nanotubes.

Of particular interest for future catalytic applications is the recent report that not only can $C_{60}$ be coated with metal atoms, but that the metal coating can consist of a precise number of metal atoms. For example, $C_{60}Li_{12}$ and $C_{60}Ca_{32}$ have been identified mass spectroscopically. $C_{60}$ has been coated with a variety of different metals, including Li, Ca, Sr, Ba, V, Ta and other transition metals. Interestingly, addition of more than 3 Ta atoms to $C_{60}$ breaks the $C_{60}$ cage. Replacement of one carbon atom in $C_{60}$ by a transition metal atom such as Co or Ir is being studied for possible catalytic applications. The future technological challenge will be to discover techniques to fabricate large quantities of such materials, so that such catalyst materials can be put in a bottle and not just in molecular beams.

Carbon nanotubes have the interesting property that they are predicted to be either semiconducting or conducting (metallic), depending on the chirality and diameter of the nanotube. Such materials are being studied as conductive additives to plastics and for use in electrochemical applications where the uniformity of the nanotube diameter and length is not overly critical. Another approach is to use the carbon nanotube as a template for a nanotube of an inorganic oxide. Hollow nanotubes of zirconia and yttria-stablilized zirconia have been prepared by coating treated carbon nanotubes with a zirconium compound and then burning out the carbon template. Finally, large scale production of single-walled nanotubes has recently been demonstrated, so one may anticipate a strong upsurge in the characterization and potential usage of single-walled carbon nanotubes in the future.

Porous carbons are of interest as molecular sieve materials, both as sorbants and as membranes, or as nanostraws for filtration. One of the major research objectives is to develop materials or structures with exceedingly high storage capacity per unit volume and weight for gases such as $H_2$ or $CH_4$. $H_2$ or $CH_4$ could become an economic source of combustion fuel or a means to power fuel cells for ultralow-emission vehicles or for electric power generation. Microporous hollow carbon fibres have exhibited high permeance and high selectivity as hydrogen selective membranes, and development is now underway to scale up these membranes to commercial levels. Carbon fibre materials produced via catalytic decomposition of hydrocarbon vapours have also recently been reported to exhibit exceptionally high hydrogen adsorption capacity. More mundane uses of nanotubes are as nanometer reinforcing rods in polymers or even in concrete. Incorporation of conducting carbon nanotubes in construction materials such as concrete or structural plastics opens opportunities for real time monitoring of material integrity and quality.

## MICROPOROUS AND DENSE ULTRATHIN FILMS

Research and development of microporous thin films for use as molecular sieving membranes using inorganic crystalline materials such as zeolites or porous silica is another area of active research around the world. For molecular sieving membranes, one critical challenge rests on discovering ways to create large scale, thin, nearly defect-free membranes. One recent example is the fabrication of mesoporous conducting thin films grown from liquid crystal mixtures. Transmission electron microscopy (TEM) reveals an ordered array of 2.5 nm diameter cylindrical holes in a 300 nm thick Pt film. The hole diameter can be varied either by changing the chain length of the surfactant molecule or by adding an alkane to the plating solution. It is interesting that this technique produces a continuous thin film with nanoscale porosity in an electrically conducting material.

Dense ultrathin films such as single monolayer films would be of significant importance in the semiconductor industry. Thin films of specialized coatings for corrosion, thermal, and/or chemical stability should be valuable for the chemical and aerospace industries. Novel chemical sensors may be anticipated through use of ultrathin films composed of specialized clusters. Typical techniques for production of thin films are physical vapour deposition, chemical vapour deposition, and Langmuir-Blodgett processes.

## MANIPULATION AT THE NANOSCALE

Over the last two decades, the development and improvement of new techniques to fabricate and characterize nanoscale materials have fueled much of the enormous growth in nanoscale science and technology, not only by making nanoscale materials relatively easily available for scientific study and characterization, but also in some instances, opening the door for large scale industrial use. For example, atomic force microscopy and scanning tunneling microscopy are two techniques that have become major workhorses for characterization of nanoscale materials. The strong upsurge in interest and funding in nanoscale materials must to some (large) degree be credited to the recent development of these two techniques. Combining X-ray structure, high resolution TEM and low energy, high resolution scanning electron microscopy (SEM), researchers now have the means to physically characterize even the smallest structures in ways impossible just a few years ago. Not only can a nanostructure be precisely examined, but its electronic character can also be mapped out.

Using the scanning probe devices, scientists can both image individual atoms and molecules and also manipulate and arrange them one at a time. This atomic manipulation to build structures is just in its infancy, but it does allow one to imagine a route to the ultimate goal of atomically tailored materials, built up atom-by-atom by a robotic synthesizer.

Of all fundamental properties controlling the stability of atoms, clusters, and particles on a surface or support, knowledge of the adsorption and adhesive energies of the metal atom or particle on the solid metal or oxide surface is critical to fundamental understanding of the stability of high surface area materials for materials applications that include oxide-supported metal catalysts, bimetallic catalysts, and metal-ceramic interfaces used in microelectronics. Knowledge of such parameters allows researchers to predict the relative strengths of the metal-metal and metal atom-support interaction energies, and to infer relative stabilities as a function of the composition and size of the metal cluster. Recently it has become possible to experimentally measure the metal atom-surface bond strength on a per-atom basis using adsorption micro-calorimetry on ultrathin single crystal metal or metal oxide surfaces. A technique such as this capable of probing interactions on an atom-by-atom or molecule-by-molecule basis can be thought of as another "atomic probe" that can be expected to substantially advance our database and understanding at the ultimate nanoscale for materials, that is, single atom binding energies to surfaces.

Similarly, new techniques are being developed to allow chemical and catalytic reactions to be followed in situ in real time. As an example, an infrared and nuclear magnetic resonance spectroscopic technique is being developed at the Max Planck Institute in Mülheim to monitor kinetics of CO adsorption on 1-3 nm diameter metal colloid particles (typically Pt, Rh, or Pd) in liquids and to follow in real time the way CO organizes itself on the particles while in liquid suspension. Such techniques will allow one to begin to understand the metal particle properties in solution and thus infer what might occur in real reaction mixtures. Extension of such techniques to real catalytic reactions in solution for catalyst particles of various sizes and composition is likely in the not too distant future.

The areas where nanoscale high surface area materials may have the greatest future impact are difficult to predict, but some signs point to the possibility of substantial advancement in the areas of adsorption/separations, particularly in gas sorption and separations and in novel chemical catalysis using nanoscale catalyst particles.

At least two major challenges must be faced before utilization and generation of high surface area nanoscale materials becomes a commonplace reality. First is critical dimensional control of the nanoscale structure over long times and varying conditions. In nanoscale catalyst materials the critical chemical selectivity is likely to be intimately associated with the local environment around what is termed the "active" site. This suggests that the size, type, and geometry of the atoms making up the active site will play a critical role in defining the conditions under which this active site will be able to carry out its designed function. Ability to fabricate materials with "exactly" the same structure and composition at each active site has been and will continue to be a major challenge to materials and catalytic scientists.

A second challenge involves thermal and chemical stability control of the fabricated nanostructure. It is generally accepted that the smaller the nanostructure (active site) the more likely it is that the structure may move, aggregate, be poisoned, decompose, or change its shape, composition or morphology upon exposure to thermal and/or chemical cycling. Identifying windows of stable operation in which the specific structure or material will be able to retain the desired (and designed) behaviour is critical for commercial applications. On the other hand, the driving force is the fact that nanostructured materials typically exhibit unique properties that are expected to open windows of opportunity previously inaccessible with existing materials.

It is important to recognize that nanoscale science and technology is not a "stand alone" field of endeavor, but rather is more of a "generic" area that is expected to have a critical impact and overlap in many areas of science and technology. The breadth of issues covered in this report can be taken as proof of this principle. The fields that fall under the "nanoscale" umbrella are many and diverse, illustrating that nanoscale science and technology is a collection of many different disciplines and areas of expertise. Such science and technology offers both an opportunity and a challenge to the scientific and technological community. Researchers in some areas of science included under the broad umbrella of nanoscale science and technology in this report do not normally consider (or in some cases, want) their scientific efforts to be labeled as "nanoscale." nanoscale science and technology broadly encompasses the science and technology that falls between that involving individual atoms or molecules at the one extreme and that involving "bulk" materials at the other extreme.

It is important to recognize that the use of nanostructuring or nanostructures to generate, fabricate or assemble high surface area materials is at an embryonic stage. The effect of the nanostructure and our ability to measure it will be increasingly important for future progress and development of materials for the marketplace. That said, it is apparent that so-called "mature" technologies such as catalysis, coatings, separations, etc., are already being impacted. Thus, one may eagerly anticipate exciting new advances in many diverse technological areas growing from our increasing understanding of nanostructuring and nanostructured materials.

## FUNCTIONAL NANOSCALE DEVICES

### Introduction

The recent emergence of fabrication tools and techniques capable of constructing structures with dimensions ranging from 0.1 to 50 nm has opened up numerous possibilities for investigating new devices in a size domain heretofore inaccessible to experimental researchers. The WTEC nanotechnology panel reviewed research in the United States, Japan, Taiwan, and Europe to find that there is considerable nanoscience and technology activity in university, industrial, and government laboratories around the world. The insight gained from this survey suggests areas of strength and areas of possible improvement in the field.

There is intense study around the world to determine the exact point in dimensional scaling where it becomes either physically unfeasible or financially impractical to continue the trend towards reducing the size while increasing the complexity of silicon chips. In some of the same laboratories where research activities on Si are decreasing, research activities on single-electron devices (SEDs) are increasing. Although there are myriad questions involving electrical contacts, interconnections, reliability, and the like, one of the fundamental issues in the miniaturization/complexity debate concerns the Si MOSFET itself when the gate length is reduced to less than 50 nm. Does it behave like a long gate device or does the output conductance increase to impractical levels due to short-channel effects?

While the signature current-voltage (I-V) characteristics provide a common basis for comparison of device performance, there are significant variations in the fabrication methods and device structures being considered by the different labs in the countries the panel surveyed that have significant SED activity. The range of research in the surveyed laboratories spans electrical measurements from millikelvin to room temperature and from discrete electronic elements to integrated single-electron transistors (SETs). Materials that are used to form the active single-electron element range from charge clusters that are shaped by electric fields in a two-dimensional electron gas to metallic colloids to single oligomers. Progress in the field is hindered by architectures based on conventional circuit approaches that fail to take sufficient advantage of the unique properties of single-charge electronics to achieve significant impact in future high density applications. Most research in SED technology is fundamental and is distributed among universities funded by government agencies.

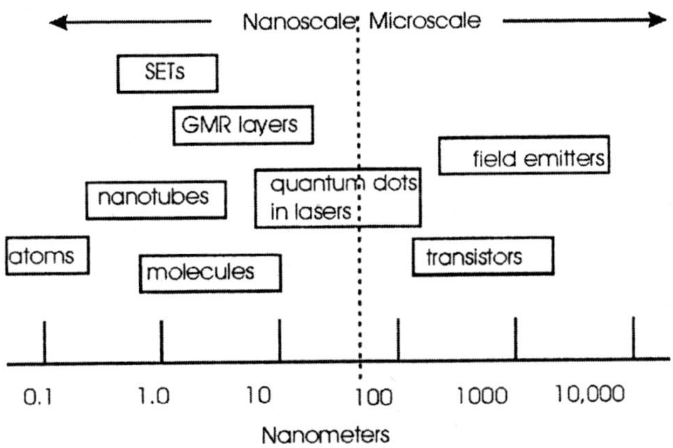


**Fig.** Functional Device Scales

The field of magnetics has experienced increasing attention since giant magnetoresistance (GMR) in multilayered structures was discovered. In these structures ferromagnetic layers are quantum mechanically coupled across a 1-3 nm nonmagnetic metallic layer. GMR structures are under intense study for applications in hard disk heads, random access memory (RAM), and sensors. Several laboratories are investigating the physics of the transition of these layers, which are quantum mechanically confined in one dimension, to layered filaments in which there are one- and two-dimensional confinements. There are numerous experimental process approaches under consideration in fabricating GMR structures, including the following:

- Magnetron or ion beam sputter deposition
- Epitaxy for layered structures
- Rubber stamping of nanoscale wire-like patterns
- Electroplating into nanoscale pores in polymer membranes

In RAM applications, a high ratio of magnetoresistance combined with a small coercive switching field is key to density, speed, and low power. These features are also achieved in magnetic tunnel junctions in which the ferromagnetic layers are quantum mechanically coupled through a thin dielectric layer. Although research in nanoscale magnetics is underway internationally, most of the activities on the practical applications mentioned above are in the United States.

Optical devices have already benefited from incorporation of nanostructured materials: commercially available semiconductor lasers incorporate active regions comprised of quantum wells, the presence of which modifies the electronic density of states and the localization of electrons and holes, resulting in more efficient laser operation. Extrapolating from those results, even greater improvements are predicted for lasers utilizing either quantum wire or quantum dot active layers. Recent advances in the "self-assembled" formation of quantum dot structures have stimulated progress in the fabrication and characterization of quantum dot lasers in Japan, Europe, and the United States.

The first synthesis and characterization of carbon nanotubes were reported. The novel material contained a wide variety of multiwalled nanotubes (MWNT) containing 2 to 50 concentric cylindrical graphene sheets with a diameter of a few nm and a length of up to 1 μm. The material was produced at the negative electrode of an arc discharge and appeared to be mixed with a large amount of other forms of carbon. This initial work led many groups throughout the world to produce and purify nanotubes. The theoretical study of their electronic structure followed in the next year. Soon it became clear that nanotubes have unique electronic and mechanical properties that are expected to lead to ground-breaking industrial applications.

## SINGLE-CHARGE ELECTRONICS

Even though the study of single-electron charging effects with granular metallic systems dates back to the 1950s, it was the research of Likharev and coworkers almost 10 years ago that laid much of the groundwork for understanding single-charge transport in nanoscale tunnel junctions. The concept was developed of a Coulomb gap that can be exploited to control the transfer of single charges. Since then, many research groups have made use of the Coulomb blockade effect to develop SED technology. Some of the myriad approaches to developing SEDs and representative laboratories pursuing the various SED concepts.

The group at Hitachi Europe uses a side-gated constriction in a delta-doped GaAs structure to fabricate a magnetic tunnel junction device in which a series of small islands separated by tunnel barriers are formed. At ~ 2 K, the Coulomb gap voltage oscillates as a function of the side-gate voltage. Using the MTJ device as a building block, both memory and logic (inverter, NOR) functions have been demonstrated. The fabrication procedure makes use of standard semiconductor processing techniques and does not rely on lithography to define the nanoscale islands, since these are created by disorder in the delta-doped layer.

Other groups have utilized fine-line lithography to fabricate SEDs. At IBM, a flash memory SED was demonstrated by fabricating a sub-50 nm Si quantum dot (QD) on top of a MOSFET channel using a silicon-on-insulator (SOI) substrate, with the QD acting as a floating gate. Single-electron charging was observed up to 90 K, while large threshold voltage shifts of up to 0.75 V were measured at 290 K. To overcome the lithography limitation on the QD size, the Toshiba group used a Si edge quantum wire approach.

An inversion layer was formed at the 15 nm high Si sidewall of a SOI structure by growing a gate oxide and depositing a poly-Si gate there. Conductance oscillations were clearly seen at 4.2 K in this edge-channel MOSFET. More recently, the Toshiba group has reverted back to a more planar device configuration, with a 50 nm wide Si quantum wire defined by e-beam lithography and oxidation of the surrounding $SiO_2$.

One method to form semiconductor QDs without depending on fine-line lithography is to make use of the self-organizing nature inherent in the Stranski-Krastanow thin film growth mode. In the initial stages of the heteroepitaxial growth of lattice-mismatched materials, strain-induced coherent relaxation occurs and dislocation-free islands are formed that are in the tens of nanometers range in size. There has been considerable research in these self-organized quantum dots (SOQDs) in the past few years, though much of the work has been of a fundamental nature.

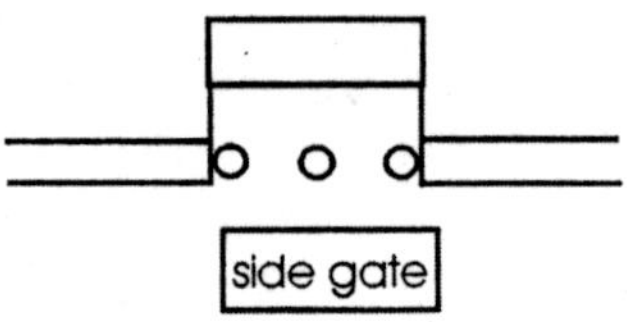


**Fig.** Metal colloids, self-assembled monolayer (SAM) coatings, polysilicon, quantum dots embedded in $SiO_2$ (Hitachi, IBM, RIKEN, NTT, ETL, University of Lund).

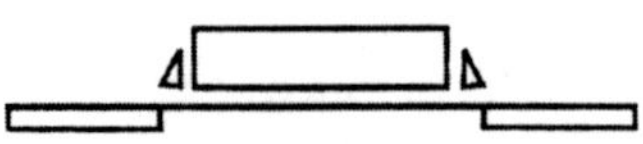

**Fig.** Sidewall extensions of MOSFET gate (Toshiba).

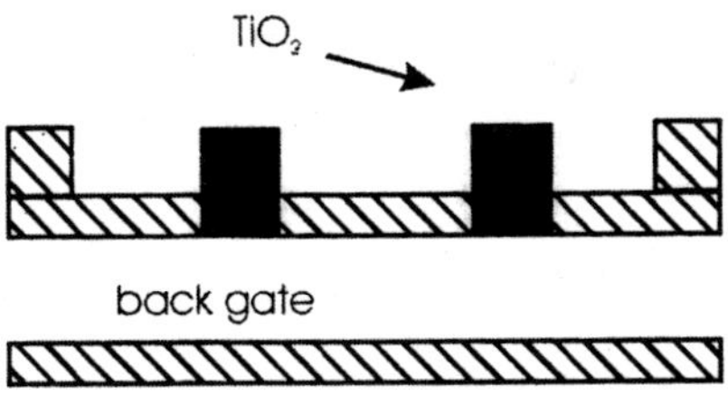


**Fig.** Oxidation of metal or semiconductor with scanning tunneling microscope (STM) tip (ETL).

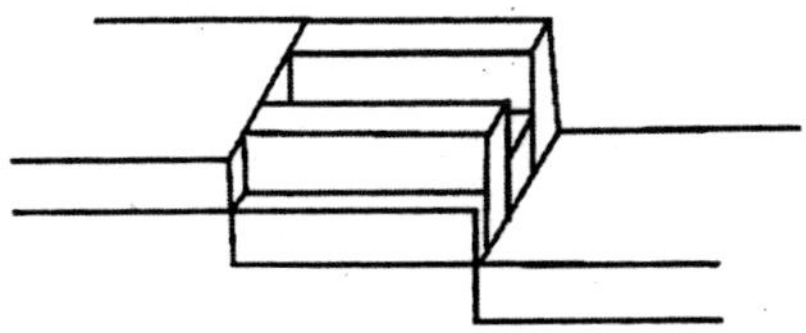

**Fig.** STM probe oxidation of metal on vicinal substrate steps (ETL).

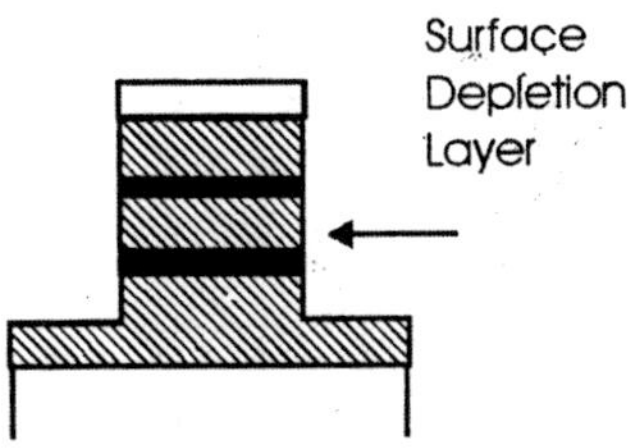


**Fig.** Double barrier tunnel diode structure.

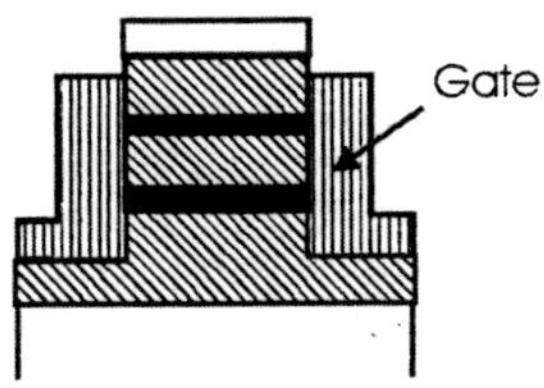


**Fig.** Gated double barrier tunnel diode structure.

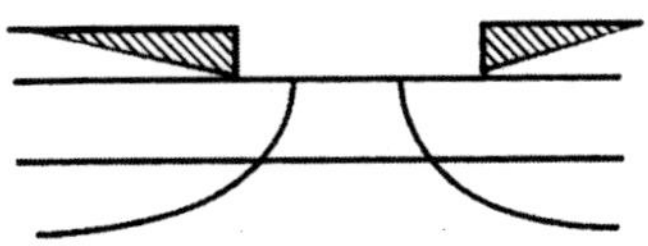

**Fig.** Depletion layer control of 2DEG area.

This concept was further demonstrated by the Sony group, which reported observing threshold voltage shifts at 300 K. The one difference in this case is that the QDs are formed at the bottom of tetrahedral-shaped recesses formed by substrate patterning. Since SOQDs form in a somewhat random manner on a planar surface, this approach provides positioning control.

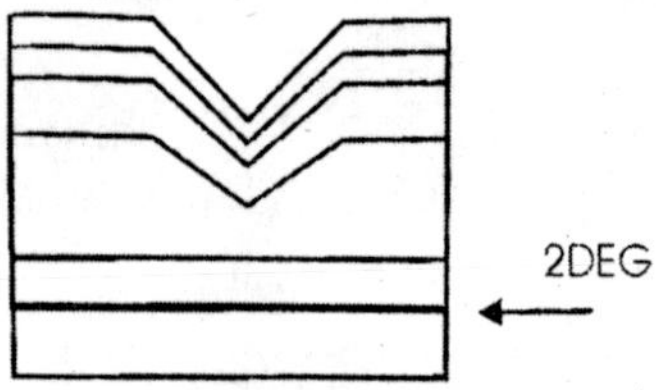


**Fig.** Tetrahedral shaped recess, TSR.

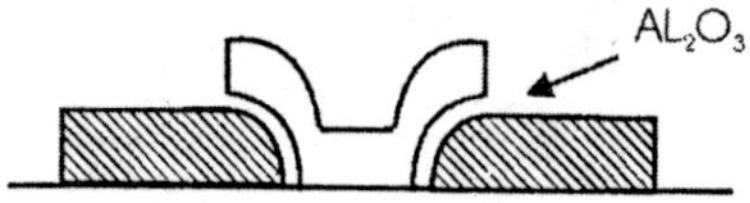


**Fig.** Double barrier metallic SET patterned by e-beam (NEC).

For a very different approach, a molecular embodiment of a QD-based system can be realised by connecting a single molecule between metallic contacts. At Yale University a single molecule of benzene-1,4-dithiol was self-assembled from solution onto two electrodes of a mechanically controllable break junction. The spacing between the electrodes is ~ 0.8 nm, and I-V measurements at room temperature showed a gap ~ 0.7 V wide, which is attributed to a Coulomb gap. The Delft University of Technology in the Netherlands is also working on transport through oligomers.

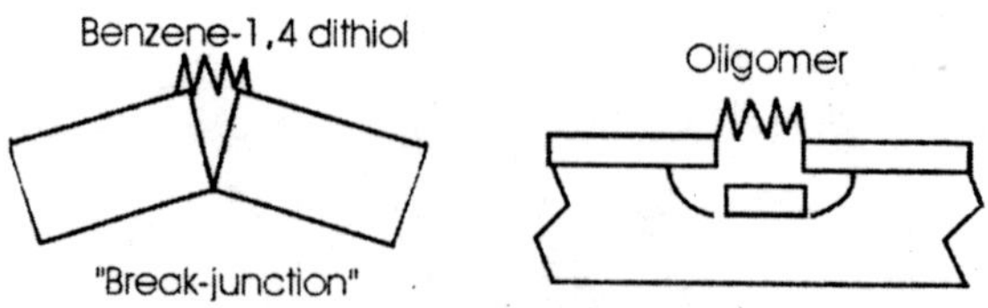


**Fig.** A single molecule connecting metallic contacts.

Concerning the architecture in which SEDs are utilized, a number of approaches have been proposed. One of the more novel ideas is that of quantum cellular automata (QCA), based on some earlier work at Texas Instruments and developed at the University of Notre Dame. The basic QCA cell is made up of a group of capacitively coupled QDs. Each cell holds two electrons, resulting in two polarization configurations that can represent the logic "0" and "1" states, and each cell interacts via Coulombic forces with neighbouring cells. An array of cells can then be used to transmit binary information, which eliminates the need for physical interconnects between devices and represents a paradigm shift for ultralarge-scale integration (ULSI). Basic Boolean operations (AND, OR, etc.) can be implemented using QCA, and more complex functions have been simulated. Most recently, the Notre Dame group has demonstrated a nonlinear, bistable response of a QCA cell, albeit at a very low temperature of less than 20 mK.

The QCA approach is not without its challenges. Circuit fabrication will be difficult because stringent control in QD positioning is required. Others have pointed out that bistability is only a necessary but not sufficient condition for the operation of Boolean logic circuits, because isolation is needed between the input and output, while background charge fluctuations will hamper logic implementation. Thus, there is also considerable research in the use of SEDs with more conventional architectures. The group at Hitachi Europe uses its MTJ devices in binary decision diagram logic that is commonly used in large-scale integrated (LSI) circuits. The Toshiba group is combining its SED with a MOSFET to compensate for the lack of gain in the former, as are, presumably, other groups working on QD-based flash memory SEDs.

The field of single-charge electronics is still limited in scope by the lack of a suitable architecture that fully utilizes the unique aspects of single-electron charging. Current approaches require many SED elements to achieve conventional functions such as adders, exclusive NORs, etc. Simulations of such circuits predict slow operating speed. The field seems to be stuck on applying conventional electronics to SEDs. Either a new architecture will be discovered or SEDs may find a niche home in those applications where measurements of single charges are needed.

**Table. Quantum Dot Flash Memory**

| | Hitachi | Minnesota | Fujitsu | IBM | Sony |
|---|---|---|---|---|---|
| Material | Poly-Si | SOI | SOI | SOI | GaAs |
| QD Material | Poly-Si | Poly-Si | Poly-Si | Poly-Si | InAs |
| QD Fab Method | E-beam/ etch | E-beam/ etch | E-beam/ etch | E-beam/ etch | Epitaxial self-assembly |
| QD Size (nm) | 10 (estim.) | 7×7×2h | 30×20×25h | 30×20×8h | 25 x 4 h |
| $2V_{th}$ (V) | 0.5–1.0 | 0.055 | 0.1 | 0.75 | 0.45 |
| Write/ Erase | 15 V/10V | >4V | 4V | 3 V | > 1 V |
| Retention | 1–24 h | 10 sec. | | > 1wk | 10 sec. |
| Circuits | 128 Mb LSI | | | | |

For the fabrication of nanoscale electronic devices, the self-organizing technique appears to be the most promising. The field of self-organized semiconductor QDs is quite active, but aside from optical emitters, very few practical electronic functions have been proposed. Of those, the single-electron flash memory is attracting attention, but there has been no serious proposal as to how the device could operate under normal integrated circuit performance conditions and reliability specifications. Using the SED as a floating gate to a MOSFET has the same kinds of problems as applying conventional approaches to SED architectures; until someone comes up with a better idea, the future of these approaches remains to be determined.

## NANOMAGNETICS

GMR in structures of alternating magnetic and nonmagnetic thin layers was the accumulation of several decades of intensive research in thin film magnetism and improvements in epitaxial growth techniques developed mainly in semiconductor materials. Not surprisingly, the first GMR structure was fabricated using molecular beam epitaxy. The high quality magnetic and nonmagnetic metallic films provide electrons with a mean free path exceeding 100 Å; on the other hand, the epitaxial growth allows for each constituent layer of the structure to be as thin as a few atomic layers. The greatly enhanced spin-dependent scattering in these multilayered structures provides magnetoresistance changes as high as 50%.

Two subsequent major developments from IBM enabled the application of GMR materials to hard disk heads, RAM, and sensors. The first development was the demonstration of equally good or better GMR materials using high throughput and production-worthy magnetron sputtering systems. The other development was the invention of magnetically soft spin-valve structures, which allow low field and low power operation.

A different class of high magnetoresistive materials was discovered in which the nonmagnetic layer separating the two ferromagnetic layers is made with an ultrathin insulating material, such as an aluminum oxide layer < 20 Å thick. With the switching of magnetization of the two magnetic layers between parallel and antiparallel states, the differences in the tunneling coefficient of the junction and thus the magnetoresistance ratio have been demonstrated to be more than 25%. A distinctive feature of this MTJ class of materials is its high impedance (> 100 kW -$\mu m^2$), which allows for large signal outputs.

The gradual improvement of GMR and MTJ materials have made them attractive for nonvolatile magnetic random access memory (MRAM) applications.

The key for a competitive MRAM technology is the fabrication of deep submicron-dimension magnetic cells. The further development of lithography tools utilizing e-beam and deep ultraviolet radiation is essential. Magnetic storage elements as small as 0.25 μm have been demonstrated by both Motorola and IBM. Among the steps of MRAM fabrication that are not yet compatible with semiconductor processing is the ion milling of the magnetic cells.

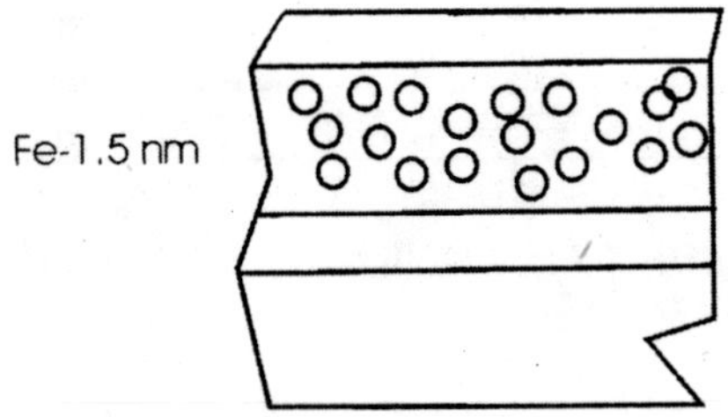


**Fig.** Granular GMR-Co, Fe.

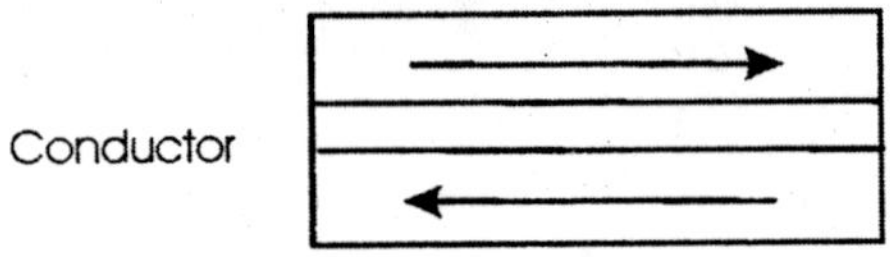


**Fig.** Current in plane.

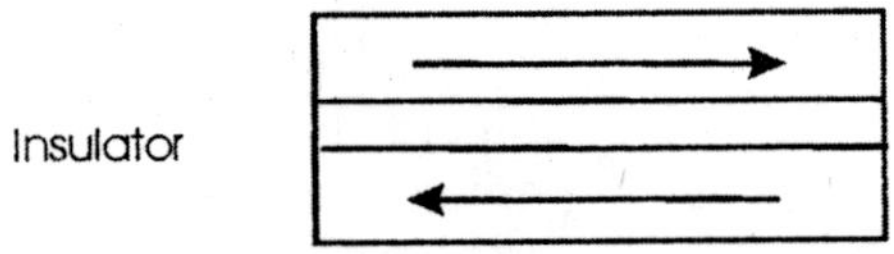


**Fig.** Magnetic tunnel junction.

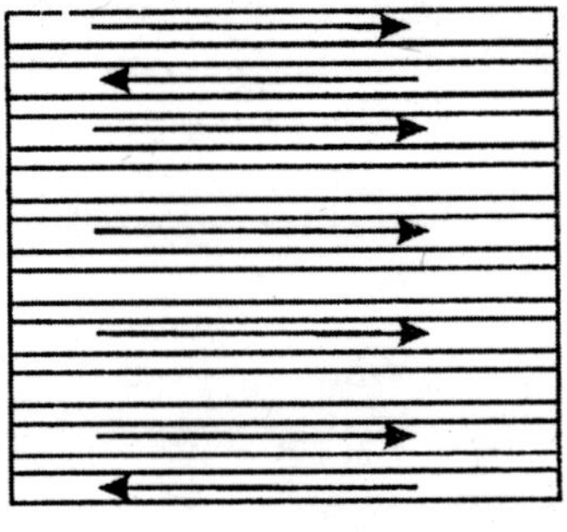

**Fig.** Ferromagnetic/metal/ferromagnetic: 3 - 60 periods free-standing (NRL, CNRS-Thomson, Philips, Michigan State, Lawrence Livermore Labs); plated into pores.

The ability to fabricate submicron magnetic elements has opened a very rich and fascinating area of micromagnetics research. Characterization techniques having nanoscale resolution have been utilized and improved to measure and image the complex magnetization patterns in order to understand the magnetization switching characteristics. Examples of such techniques include the following:

- Superconducting quantum interference device (SQUID) magnetometry
- Magnetic force microscopy (MFM)
- Scanning electron microscopy with polarization analysis (SEMPA)
- Magnetic near-field microscopy
- Electron holography

Another essential tool is micromagnetics modeling, which is used to predict complex magnetic domain configurations in patterned magnetic elements and to generate transient pictures that demonstrate the process of forming complex domain configurations.

By combining MFM, SQUID magnetometry, SEMPA, and micromagnetics modeling, researchers at Motorola have conducted a systematic study of the switching characteristics of single-layer and multilayer submicron magnetic structures. Three different phases of the magnetization phase diagram have been identified with regards to material composition, dimension, shape, thickness, and other structural parameters:

1. The quasi-single domain phase can be well described by the coherent rotation model,
2. The end-domain phase is dominated by the nonuniform regions of magnetization at the two ends of the element-the magnetization switching process can either be rotational or through domain wall nucleation and propagation, and
3. The trapped magnetization vortice (TMV) phase, which is characterized by the presence of magnetization vortices. Nucleation from TMV sites requires lower reversal fields than coherent rotation, but a high field is needed to drive out TMVs in the element. When the driving field is not high enough, TMVs can cause unusually large fluctuations in the switching field.

As the size of magnetic elements scales below 20 nm, a superparamagnetic phase emerges in which the room temperature thermal energy overcomes the magnetostatic energy well of the element, resulting in zero hysteresis. In other words, although the element itself is a single-domain ferromagnet, the ability of an individual magnetic "dot" to store magnetization orientation information is lost when its dimension is below a threshold. On the other hand, suitably prepared alloys of immiscible ferromagnetic and nonmagnetic metals that contain single-domain ferromagnetic grains in a nonmagnetic matrix have been shown to exhibit GMR characteristics. The moments of the magnetic grains are aligned at high fields and random at the coercive field, leading to GMR characteristics. In such "granular" metals, GMR has been reported for sputtered alloy films of CoCu, FeCu, NiFe/Ag, and CoAg. GMR values as high as 55% at 4.2 K and 20% at room temperature have been observed. The granular films require magnetic fields of the order of 10 kOe to achieve such a change in electrical resistance.

A very exciting consequence of ultrascaled magnetic particles is quantum tunneling of the magnetization direction of a collection of spins. There is no simple Schrödinger equation that describes this process, since it is not an elementary particle that is tunneling but a collective coordinate. Below its "blocking temperature," at which thermally assisted hopping between magnetic orientations becomes frozen out, magnetic particles of TbCeFe at sizes around 15 nm have been observed to behave independent of temperature and with no freeze-out magnetic relaxation. Because of the coherent tunneling of the magnetization orientation between the symmetric double-well potential, a resonance line in the magnetic susceptibility and noise spectra has been observed at temperatures below 200 mK in zero applied magnetic field. This work has stimulated a number of theoretical investigations into the effects of dissipation and the feasibility of producing quantum effects in larger magnetic structures.

Another interesting type of nanomagnetic structure is nanometer ferromagnetic wires fabricated using conventional nanolithography, nanoimprint lithography, AFM/MFM direct writing, groove deposition, and electrodeposition into pores of template polymer membranes. Such nanowires of either single layer or multilayers may provide new approaches to very small magnetoresistive sensors, ultrahigh-density hard disks, and other extensions of conventional applications. Another intriguing possibility is the suggestion to use heterostructure nanowires to investigate single electron tunneling.

Recently, molecular magnetism has received much attention with the development of a variety of synthesis techniques largely adapted from biology and chemistry. Natural and artificial ferritin proteins are examples of systems obtained using these methods. The ability to add one magnetic ion at a time has resulted in nanoscale magnets precisely defined by atomic weight. Ferritin consists of a segmented protein shell in the shape of a hollow sphere, with an outer diameter of 12.5 nm and an inner diameter of 7.5 nm. In vivo, the inner space is normally filled with a crystal of an iron oxide that is antiferromagnetic below 240 K. The empty protein shells can also serve as vessels for the synthesis of ferrimagnetic magnetite and maghemite. Thus, there exists a system in which its size as well as the nature of its magnetic interactions can be varied.

Another example of molecular magnetism is a cobalt-iron-cyanide-based Prussian blue analog. In the ground state the $Fe^{+2}$ and $Co^{+3}$ ions are low-spin and diamagnetic, and there is no interaction between them. Red light excitation transfers one electron from an iron site to a cobalt site, resulting in high-spin $Fe^{+3}$ and $Co^{+2}$ ions and magnetic interactions between them. The application of a blue light causes a transition back to the initial state and switches off the $Fe^{+3}$-$Co^{+2}$ interactions. In addition to this kind of photochemically controllable magnets, electrochemically controllable magnets have also been reported.

## QUANTUM DOT LASERS

Semiconductor lasers are key components in a host of widely used technological products, including compact disk players and laser printers, and they will play critical roles in optical communication schemes. The basis of laser operation depends on the creation of nonequilibrium populations of electrons and holes, and coupling of electrons and holes to an optical field, which will stimulate radiative emission. Calculations carried out in the early 1970s predicted the advantages of using quantum wells as the active layer in such lasers: the carrier confinement and nature of the electronic density of states should result in more efficient devices operating at lower threshold currents than lasers with "bulk" active layers. In addition, the use of a quantum well, with discrete transition energy levels dependent on the quantum well dimensions (thickness), provides a means of "tuning" the resulting wavelength of the material. The critical feature size-in this case, the thickness of the quantum well-depends on the desired spacing between energy levels. For energy levels of greater than a few tens of millielectron volts (meV, to be compared with room temperature thermal energy of 25 meV), the critical dimension is approximately a few hundred angstroms. Although the first quantum well laser, demonstrated, was many times less efficient than a conventional laser, through the use of new materials growth capabilities (molecular beam epitaxy), and optimization of the heterostructure laser design.

Even greater benefits have been predicted for lasers with quantum dot active layers. In the early 1980s that quantum dot lasers should exhibit performance that is less temperature-dependent than existing semiconductor lasers, and that will in particular not degrade at elevated temperatures. Other benefits of quantum dot active layers include further reduction in threshold currents and an increase in differential gain-that is, more efficient laser operation. Stimulated recombination of electron-hole pairs takes place in the GaAs quantum well region, where the confinement of carriers and of the optical mode enhance the interaction between carriers and radiation. In particular, note the change in the electronic density of states, as a function of the "dimensionality" of the active layer. The population inversion (creation of electrons and holes) necessary for lasing occurs more efficiently as the active layer material is scaled down from bulk (3-dimensional) to quantum dots (0-dimensional). However, the advantages in operation depend not only on the absolute size of the nanostructures in the active region, but also on the uniformity of size. A broad distribution of sizes "smears" the density of states, producing behaviour similar to that of bulk material.

Thus, the challenge in realizing quantum dot lasers with operation superior to that shown by quantum well lasers is that of forming high quality, uniform quantum dots in the active layer. Initially, the most widely followed approach to forming quantum dots was through electron beam lithography of suitably small featured patterns (~300 Å) and subsequent dry-etch transfer of dots into the substrate material. The problem that plagued these quantum dot arrays was their exceedingly low optical efficiency: high surface-to-volume ratios of these nanostructures and associated high surface recombination rates, together with damage introduced during the fabrication itself, precluded the successful formation of a quantum dot laser.

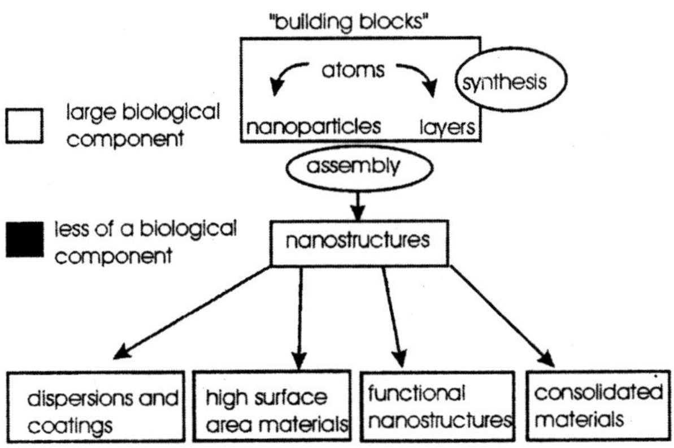


**Fig.** Schematic of a semiconductor laser

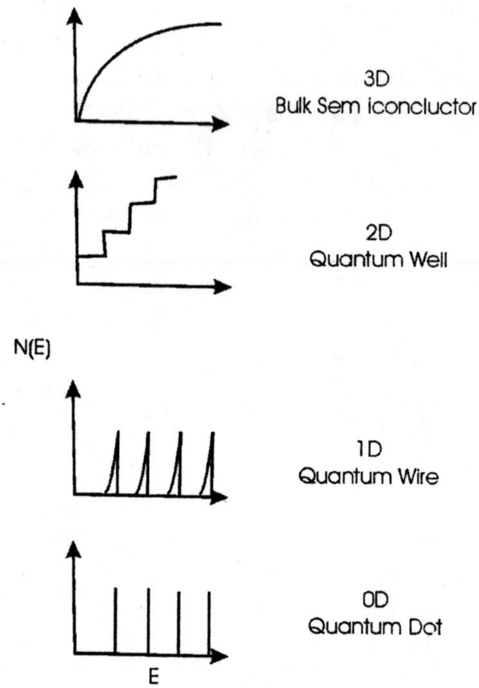


**Fig.** Density of electronic states as a function of structure size

With the demonstration of the high optical efficiency self-assembled formation of quantum dots, formed without need of external processing and having the natural overgrowth of cladding material there ensued a marked increase in quantum dot laser research. The first demonstration of a quantum dot laser with high threshold density. The density of the quantum dot structures, stacking successive, strain-aligned rows of quantum dots and therefore achieving vertical as well as lateral coupling of the quantum dots. In addition to utilizing their quantum size effects in edge-emitting lasers, self-assembled quantum dots have also been incorporated within vertical cavity surface-emitting lasers. Table gives a partial summary of the work and achievements in quantum dot lasers.

As with the demonstration of the advantages of the quantum well laser that preceded it, the full promise of the quantum dot laser must await advances in the understanding of the materials growth and optimization of the laser structure. Although the self-assembled dots have provided an enormous stimulus to work in this field, there remain a number of critical issues involving their growth and formation: greater uniformity of size, controllable achievement of higher quantum dot density, and closer dot-to-dot interaction range will further improve laser performance. Better understanding of carrier confinement dynamics and capture times, and better evaluation of loss mechanisms, will further improve device characteristics.

It should be noted that the spatial localization of carriers brought about by the quantum dot confinement may play a role in the "anomalous" optical efficiency of the GaN-based materials, which is exceptional in light of the high concentration of threading dislocations (~ $10^8$ - $10^{10}$ $cm^{-2}$) that currently plague this material system. The localization imposed by the perhaps natural nanostructure of the GaN materials may make the dislocation largely irrelevant to the purely optical (but not to the electrical) behaviour of the material.

## CARBON NANOTUBES

The first synthesis and characterization of carbon nanotubes were reported by Iijima from NEC in late nineties. The initial theoretical study of their electronic structure was soon followed with the work by Dresselhaus and coworkers at MIT. Since then, the fabrication of nanotubes has been improved by several groups, and methods other than arc discharge have been explored. The main issues are to separate the nanotubes from other forms of carbon also produced in the fabrication process and to increase the yield of single-walled nanotubes (SWNT) for potential applications. Following Iijima's work, macroscopic quantities of MWNT were produced with an improved arc discharge method by Ebbesen and coworkers at NEC.

SWNT can be efficiently produced by laser ablation of a graphite rod. In the following year, that same group produced what is considered to be among the best SWNT material generated so far; over 70% of the volume of material was nanotubes bundled together into crystalline ropes of metallic character. A group from the Chinese Academy of Science used chemical vapour deposition to produce a 50 mm thick film of nanotubes that were highly aligned perpendicular to the surface. Progress in recent years leads one to predict that it will indeed be possible to produce high quality carbon nanotubes in macroscopic quantities needed for many of the applications outlined below.

Nanotube bundles form a low density material and are expected to have high stiffness and axial strength as a result of their seamless cylindrical graphitic structure. It is therefore predicted that they can be used to fabricate a material with better mechanical properties than the present carbon fibre materials. Information about the mechanical properties of nanotubes has been gathered recently by a study of the thermal vibrations of a single SWNT attached to a substrate. Ebbesen's group at the Princeton NEC Research Institute found that nanotubes have an exceptionally high Young's modulus ($\sim 2 \times 10^9$ Pa). In order to reach a better understanding of the mechanical properties and intrinsic limitations of nanotubes, Bernholc's group from North Carolina State University theoretically studied the behaviour of nanotubes beyond the linear Hooke's law and the nature of the defects leading to dislocations and fractures.

Nanotubes are highly polarizable nanoscale straws, a property that confers on them the capacity to ingest inorganic elements by nanocapillarity. As a result, it has been conjectured that they could be used as minute molds to shape nanometer-sized quantum wires and as miniature test tubes. The efficiency of their process is low, and prior removal of the caps from the ends of the nanotubes is expected to improve the situation.

Metallic nanotubes are strongly polarizable in an electric field and thereby lead to field enhancement at their extremity, the strength of which depends on the ratio of the diameter to the length and can be extremely large for routinely produced nanotubes. For this reason and possibly others related to quantum confinement effects, nanotubes are expected to form outstanding field-emitting materials.

## BULK BEHAVIOUR OF NANOSTRUCTURED MATERIALS

Bulk nanostructured materials are defined as bulk solids with nanoscale or partly nanoscale microstructures. This category of nanostructured materials has historical roots going back many decades but has a relatively recent focus due to new discoveries of unique properties of some nanoscale materials.

Early in the century, when "microstructures" were revealed primarily with the optical microscope, it was recognized that refined microstructures, for example, small grain sizes, often provided attractive properties such as increased strength and toughness in structural materials. A classic example of property enhancement due to a refined microstructure-with features too small to resolve with the optical microscope-was age hardening of aluminum alloys. The phenomenon, discovered by Alfred Wilm in 1906, was essentially explained by Merica, Waltenberg, and Scott in 1919, and the microstructural features responsible were first inferred by the X-ray studies of Guinier and Preston in 1938. With the advent of transmission electron microscopy (TEM) and sophisticated X-ray diffraction methods it is now known that the fine precipitates responsible for age hardening, in Al-4% Cu alloys, for example, are clusters of Cu atoms-Guinier-Preston (GP) zones-and the metastable partially coherent q ′ precipitate. Maximum hardness is observed with a mixture of GPII (or q ″) (coarsened GP zones) and q ′ with the dimensions of the q ′ plates, typically about 10 nm in thickness by 100 nm in diameter. Therefore, the important microstructural feature of age-hardened aluminum alloys is nanoscale. There are a number of other examples of nanoscale microstructures providing optimized properties. The critical current density $J_C$ of commercial superconducting $Nb_3Sn$ is controlled by grain size and is inversely proportional to grain size, with grain sizes of 50-80 nm providing high values of $J_{C.}$

The field of nanocrystalline (or nanostructured, or nanophase) materials as a major identifiable activity in modern materials science results to a large degree from the work in the 1980s of Gleiter and coworkers, who synthesized ultrafine-grained materials by the in situ consolidation of nanoscale atomic clusters. The ultrasmall size (< 100 nm) of the grains in these nanocrystalline materials can result in dramatically improved-or different-properties from conventional grain-size (> 1 μm) polycrystalline or single crystal materials of the same chemical composition. This is the stimulus for the tremendous appeal of these materials.

While there are a number of bulk properties that may be dramatically changed when the microstructure is nanoscale,

These are:

1. The mechanical properties of nanostructured materials for a variety of potential structural applications, and
2. Ferromagnetic materials with nanoscale microstructures for potential applications as soft magnetic materials and permanent magnet materials, and for other special applications such as information storage, magnetoresistance spin valves, and magnetic nanocomposite refrigerants.

## STRUCTURAL NANOSTRUCTURED MATERIALS

The great interest in the mechanical behaviour of nanostructured materials originates from the unique mechanical properties first observed and/or predicted for the materials prepared by the gas condensation method. Among these early observations/predictions were the following:

- Lower elastic moduli than for conventional grain size materials-by as much as 30 - 50%
- Very high hardness and strength-hardness values for nanocrystalline pure metals (~ 10 nm grain size) are 2 to 7 times higher than those of larger grained (>1 m m) metals

- A negative Hall-Petch slope, i.e., decreasing hardness with decreasing grain size in the nanoscale grain size regime
- Ductility-perhaps superplastic behaviour-at low homologous temperatures in brittle ceramics or intermetallics with nanoscale grain sizes, believed due to diffusional deformation mechanisms

While some of these early observations have been verified by subsequent studies, some have been found to be due to high porosity in the early bulk samples or to other artifacts introduced by the processing procedures.

## ELASTIC PROPERTIES

Early measurements of the elastic constants on nanocrystalline (nc) materials prepared by the inert gas condensation method gave values, for example for Young's Modulus, *E*, that were significantly lower than values for conventional grain size materials. While various reasons were given for the lower values of *E*, it was suggested by Krstic and coworkers that the presence of extrinsic defects-pores and cracks, for example-was responsible for the low values of *E* in nc materials compacted from powders. This conclusion was based on the observation that nc NiP produced by electroplating with negligible porosity levels had an *E* value comparable to fully dense conventional grain size Ni. Subsequent work on porosity-free materials has supported these conclusions, and it is now believed that the intrinsic elastic moduli of nanostructured materials are essentially the same as those for conventional grain size materials until the grain size becomes very small, e.g., < 5 nm, such that the number of atoms associated with the grain boundaries and triple junctions becomes very large. Nanocrystalline Fe prepared by mechanical attrition and measured by a nano-indentation technique. Thus, for most nanostructured materials (grain size > 10 nm), the elastic moduli are not unique properties and not a "negative."

## HARDNESS AND STRENGTH

Hardness and strength of conventional grain size materials (grain diameter, d > 1 mm) is a function of grain size. For ductile polycrystalline materials the empirical Hall-Petch equation has been found to express the grain-size dependence of flow stress at any plastic strain out to ductile fracture. In terms of yield stress, this expression is s o = $s_i$ + $kd^{-1/2}$, where s o = yield stress, $s_i$ = friction stress opposing dislocation motion, k = constant, and d = grain diameter. Similar results are obtained for hardness, with Ho = $H_i$ + $kd^{-1/2}$. To explain these empirical observations, several models have been proposed, which involve either dislocation pileups at grain boundaries or grain boundary dislocation networks as dislocation sources. In all cases the Hall-Petch effect is due to dislocation motion/generation in materials that exhibit plastic deformation.

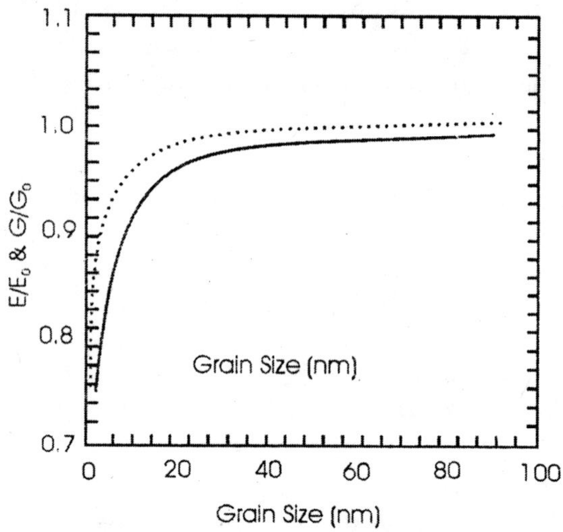


**Fig.** Ratio of the Young's (*E*) and shear (G) moduli of nanocrystalline materials to those of conventional grain size materials as a function of grain size. The dashed and solid curves correspond to a grain boundary thickness of 0.5 and 1 nm, respectively.

Most of the mechanical property data on nc materials have pertained to hardness, although some tensile test data are becoming available. Several recent reviews have summarized the mechanical behaviour of these materials. It is clear that as grain size is reduced through the nanoscale regime (< 100 nm), hardness typically increases with decreasing grain size and can be factors of 2 to 7 times harder for pure nc metals (10 nm grain size) than for large-grained (> 1 μm) metals.

The experimental results of hardness measurements, summarized previously, show different behaviour for dependence on grain size at the smallest nc grains (< 20 nm), including (a) a positive slope ("normal" Hall-Petch behaviour), (b) essentially no dependence (~ zero slope), and (c) in some cases, a negative slope. Most data that exhibit the negative Hall-Petch effect at the smallest grain sizes have resulted from nc samples that have been annealed to increase their grain size. It is suggested that thermally treating nanophase samples in the as-produced condition may result in such changes in structure as densification, stress relief, phase transformations, or grain boundary structure, all of which may lead to the observed negative Hall-Petch behaviour. Only a few cases of negative Hall-Petch behaviour have been reported for as-produced nanocrystalline samples with a range of grain sizes. These include electrodeposited nc alloys and devitrified nc alloys. Nanocrystalline thin films with grain sizes £ 6 nm are also observed to exhibit a negative Hall-Petch effect. While it seems likely that in many cases the observed negative Hall-Petch slopes are due to artifacts of the specimen preparation methods, it is also likely that conventional dislocation-based deformation is not operable in nanocrystalline materials at the smallest grain sizes (< ~30 nm). At these grain sizes, theoretically, mobile dislocations are unlikely to occur; nor have they been observed in *in situ* TEM deformation experiments. Thus, the hardness, strength, and deformation behaviour of nanocrystalline materials is unique and not yet well understood.

## DUCTILITY AND TOUGHNESS

It is well known that grain size has a strong effect on the ductility and toughness of conventional grain size (> 1 mm) materials. For example, the ductile/brittle transition temperature in mild steel can be lowered about 40°C by reducing the grain size by a factor of 5. On a very basic level, mechanical failure, which limits ductility, is an interplay or competition between dislocations and cracks. Nucleation and propagation of cracks can be used as the explanation for the fracture stress dependence on grain size. Grain size refinement can make crack propagation more difficult and therefore, in conventional grain size materials, increase the apparent fracture toughness. However, the large increases in yield stress (hardness) observed in nc materials suggest that fracture stress can be lower than yield stress and therefore result in reduced ductility. The results of ductility measurements on nc metals are mixed and are sensitive to flaws and porosity, surface finish, and method of testing (e.g., tension or compression testing). In tension, for grain sizes < 30 nm, essentially brittle behaviour has been observed for pure nanocrystalline metals that exhibit significant ductility when the grain size is conventional.

In some metals, Cu for example, ductile behaviour is observed in compression, along with yield strengths about twice those observed in tension. While it is likely that the flaws and porosity present in many nc samples seriously affect the results of mechanical tests and may be partly responsible for the asymmetry of results in compression compared to tension tests, the nature of the deformation process in terms of shear banding may also be important. The above behaviour is presumably due to the inability of usual dislocation generation and motion to occur at these smallest nc grain sizes.

An intriguing suggestion based on early observations of ductile behaviour of brittle nc ceramics at low temperatures is that brittle ceramics or intermetallics might exhibit ductility with nc grain structures. Apparent plastic behaviour in compression in nc $CaF_2$ at 80°C and nc $TiO_2$ at 180°C. These observations were attributed to enhanced diffusional creep providing the plasticity at these temperatures, where conventional grain-size materials would fail in the elastic regime. It was assumed that diffusional creep was responsible for the plasticity; observations were rationalized, with boundary diffusion dominating the behaviour such that the strain (creep) rate is defined as

$$\frac{d\varepsilon}{dt} = \frac{B\sigma\Omega\Delta D_b}{d^3kT}$$

where σ is the applied stress, Ω the atomic volume, d the grain size, k the Boltzmann constant, T the temperature, B a constant, and $D_b$ the grain boundary diffusion coefficients. Going from a grain size of 1 μm to 10 nm should increase dε / dt by $10^6$ or more if $D_b$ is significantly larger for nc materials. However, these results on nc $CaF_2$ and nc $TiO_2$ have not been reproduced, and it is believed that the porous nature of these samples was responsible for the apparent ductile behaviour. In addition, the idea of unusually high creep rates at low temperatures has been refuted. Recent creep measurements of nc Cu, Pd, and Al-Zr at moderate temperatures find creep rates comparable to or lower than corresponding coarse-grain rates. The creep curves at low and moderate homologous temperatures could be fit by the equation for exhaustion (logarithmic) creep. One explanation is that the observed low creep rates are caused by the high fraction of low energy grain boundaries in conjunction with the limitation on dislocation activity by the small grain sizes.

In sum, the predicted ductility due to diffusional creep in nc brittle ceramics or intermetallics at temperatures significantly less than 0.5 $T_M$ has not been realised.

## SUPERPLASTIC BEHAVIOUR

Superplasticity is the capability of some polycrystalline materials to exhibit very large tensile deformations without necking or fracture. Typically, elongations of 100% to > 1000% are considered the defining features of this phenomenon. As grain size is decreased it is found that the temperature is lowered at which superplasticity occurs, and the strain rate for its occurrence is increased. Equation suggests that creep rates might be enhanced by many orders of magnitude and superplastic behaviour might be observed in nc materials at temperatures much lower than 0.5 $T_M$. As mentioned above, actual creep experiments have not borne out this prediction, but instead have shown creep rates comparable to or lower than those in coarse-grained samples of the same material. This is presumably why little enhancement in ductility or superplastic behaviour has been observed for nc materials at temperatures < 0.5 $T_M$. However, there is evidence of enhancement of superplastic behaviour in nc materials at temperatures > 0.5 $T_M$. Superplasticity has been observed at somewhat lower temperatures and at higher strain rates in nc materials. The evidence for tensile superplasticity is limited and observed typically at temperatures greater than 0.5 $T_M$ and in materials that exhibit superplasticity in coarser grain sizes (1-10 μm). Superplastic behaviour in nc Pb-62%Sn at 0.64 $T_M$ and nc Zn-22%Al at 0.52 to 0.60 $T_M$. However, superplasticity (190% elongation, m = 0.32) was observed in Ti at 0.42 $T_M$. This was at a temperature 50°C lower than for 10 μm grain size Ti. The flow stress for the 200 nm Ti at 550°C was 90 MPa, compared to 120 MPa for 10 μm Ti at 600°C.

Superplastic behaviour in $Ni_3Al$ with a 50 nm grain size at temperatures of 0.56 to 0.60 $T_M$ to strains of 300 - 600%, but with unusual stress-strain behaviour and significant apparent strain-hardening. These new results suggest very different mechanisms may be causing superplastic behaviour in these nc materials.

## UNIQUE MECHANICAL PROPERTIES OF NANOCRYSTALLINE MATERIALS

While there are still only limited data on the mechanical behaviour-especially tensile properties-of nc materials, some generalizations may be made regarding the deformation mechanisms. It is likely that for the larger end of the nanoscale grain sizes, about 50 - 100 nm, dislocation activity dominates for test temperatures < 0.5 $T_M$. As grain size decreases, dislocation activity apparently decreases. The essential lack of dislocations at grain sizes below 50 nm is presumably the result of the image forces that act on dislocations near surfaces or interfaces. The lack of dislocations in small, confined spaces such as single-crystal whiskers has been known for many years. Creation of new dislocations is also made difficult as the grain size reaches the lower end of the nanoscale (< 10 nm). Stresses needed to activate dislocation sources, such as the Frank-Read source, are inversely proportional to the distance between dislocation pinning points. Since nanoscale grains will limit the distance between such pinning points, the stresses to activate dislocation sources can reach the theoretical shear stress of a dislocation-free crystal at the smallest grain sizes (~ 2 nm). Thus, at the smallest grain sizes we may have new phenomena controlling deformation behaviour. It has been suggested that such phenomena may involve grain boundary sliding and/or grain rotation accompanied by short-range diffusion-assisted healing events.

Several examples of deformation by shear banding have been reported for nc materials. In all cases, deformation in compression proceeds by intense localized shear banding. The stress-strain curves exhibited essentially elastic, perfectly plastic behaviour; that is, no measurable strain hardening was observed. Shear banding is also the deformation mode observed in amorphous metallic alloys and amorphous polymers. The deformation shear banding in nc Fe-10% Cu was compared to that for metallic glasses, amorphous polymers, and coarse-grained polycrystalline metals after significant plasticity and work hardening had taken place. While this suggests a close similarity between deformation in nc materials and amorphous materials, not all tensile data on nc materials exhibit a lack of strain hardening. The Fe-10% Cu samples shear bands even in their larger grained specimens (i.e., about 1,000 nm).

## UNIQUE MECHANICAL PROPERTIES OF NANOCRYSTALLINE MATERIALS

While there are still only limited data on the mechanical behaviour-especially tensile properties-of nc materials, some generalizations may be made regarding the deformation mechanisms. It is likely that for the larger end of the nanoscale grain sizes, about 50 - 100 nm, dislocation activity dominates for test temperatures < 0.5 $T_M$. As grain size decreases, dislocation activity apparently decreases. The essential lack of dislocations at grain sizes below 50 nm is presumably the result of the image forces that act on dislocations near surfaces or interfaces. The lack of dislocations in small, confined spaces such as single-crystal whiskers has been known for many years. Creation of new dislocations is also made difficult as the grain size reaches the lower end of the nanoscale (< 10 nm). Stresses needed to activate dislocation sources, such as the Frank-Read source, are inversely proportional to the distance between dislocation pinning points. Since nanoscale grains will limit the distance between such pinning points, the stresses to activate dislocation sources can reach the theoretical shear stress of a dislocation-free crystal at the smallest grain sizes (~ 2 nm). Thus, at the smallest grain sizes we may have new phenomena controlling deformation behaviour. It has been suggested that such phenomena may involve grain boundary sliding and/or grain rotation accompanied by short-range diffusion-assisted healing events.

Several examples of deformation by shear banding have been reported for nc materials.In all cases, deformation in compression proceeds by intense localized shear banding. The stress-strain curves exhibited essentially elastic, perfectly plastic behaviour; that is, no measurable strain hardening was observed. Shear banding is also the deformation mode observed in amorphous metallic alloys and amorphous polymers. The deformation shear banding in nc Fe-10% Cu was compared to that for metallic glasses, amorphous polymers, and coarse-grained polycrystalline metals after significant plasticity and work hardening had taken place. While this suggests a close similarity between deformation in nc materials and amorphous materials, not all tensile data on nc materials exhibit a lack of strain hardening.

## THEORETICAL NEEDS

The lack of understanding of the microscopic deformation and fracture mechanisms in nc materials. Clearly, a stronger theoretical effort is needed to guide critical experiments and point the direction for optimizing properties. There has been limited work in this area, especially in Russia, in applying disclination theory to grain rotation. Another alternate deformation mechanism that may be important in nc materials is mechanical twinning. Little theoretical work has been carried out to address this possibility. Another important potential approach to understanding the deformation mechanisms in nc materials is to explore the rich older literature on shear banding-mechanical instabilities that have been observed in, for example, mild steels as grain size decreases; the approach to very low strain hardening is observed. Amorphous materials exhibit many of the phenomenological characteristics of deformation in nc materials, that is, shear banding, asymmetry between tensile and compressive behaviour, and perfectly plastic behaviour. However, deformation mechanisms are not well understood in amorphous materials either. Recent work on bulk metallic glasses may help clarify this.

## APPLICATIONS

Of the present or near-term applications for nc materials, the hard material WC/Co is an example of several important trends. Nanostructured WC/Co composites have been prepared that can have the following characteristics:

- Nanoscale grain sizes after consolidation of powder
- Hardness values about twice that of conventional micrograined WC/Co
- Enhanced wear resistance and cutting performance

While this material has increased hardness and strength, preliminary reports from Stevens Institute of Technology in Hoboken, NJ, and the Royal Institute of Technology in Stockholm, Sweden, point to similar or increased fracture toughness values for nanostructured WC/Co. As noted earlier, single-phase nanostructured materials studied to date have exhibited high strength and hardness but brittle behaviour at low homologous temperatures ($< 0.5\ T_M$). The results of these studies of WC/Co two-phase nanostructured materials suggest that the combination of high hardness/strength and toughness/ductility may be possible in multiphase nanostructured materials.

Other examples that point to this possibility come from the work of Professor A. Inoue at Tohoku University in Japan, whose lab the WTEC panel visited. Inoue and coworkers have synthesized a variety of multiphase Al, Mg, and Ni-base alloys with nanoscale microstructures. Many of these alloys consist of nanocrystallites in an amorphous matrix. Some Al-rich alloys contain nanoscale quasi-crystalline particles surrounded by crystalline face-centreed cubic Al. The fascinating properties of these multiphase nanostructured alloys include extremely high strength coupled with some ductility. Ductility is high in compression, but uniform elongation in tension is limited. Again, this behaviour is analogous to that exhibited by ductile amorphous alloys. These results again suggest the possibility for development of nanostructured multiphase composites that combine extremely high hardness and strength with toughness and ductility. Such materials could have many applications as unique structural materials.

Multiphase ceramic nanocomposites are the focus of several efforts. One significant effort is underway at the Research Centre for Intermaterials of the Institute of Scientific and Industrial Research at Osaka University in Japan. Professor Koichi Niihara has a large effort studying micro-nano composites such as $Al_2O_3$/SiC that have enhanced toughness and also a variety of hard matrix/soft dispersion or soft matrix/hard dispersion nanocomposites.

While "low temperature" superplasticity has not been realised in nanostructured materials, enhanced superplastic behaviour has been observed at elevated temperatures in terms of a somewhat lower temperature range for superplasticity and, perhaps more significant, a higher strain rate regime. While much needs to be understood about superplastic behaviour of nanostructured materials, the possibility for using more conventional strain rates for forming has major industrial implications. This is of particular interest for near net shape forming of ceramics.

## FERROMAGNETIC NANOSTRUCTURED MATERIALS

The discovery of nanocrystalline Fe-based soft magnetic materials is less than ten years old. The first class of such materials was the melt-spun Fe-Si-B alloys containing small amounts of Nb and Cu. The Fe-Si-B-Nb-Cu amorphous phase transforms to a body-centreed cubic (bcc) Fe-Si solid solution with grain sizes of about 10 nm during annealing at temperatures above the crystallization temperature. The presence of small amounts of Cu helps increase the nucleation rate of the bcc phase while Nb retards the grain growth. These "Finemet" alloys provide low core losses (even lower than amorphous soft magnetic alloys such as Co-Fe-Si-B), exhibit saturation induction of about 1.2 T, and exhibit very good properties at high frequencies, comparable to the best Co-based amorphous alloys. These were first developed in Japan and have stimulated a large amount of research and development worldwide to optimize the magnetic properties. There has been relatively little work in the United States in this area, however.

While many of the soft magnetic properties of Finemet-type nanocrystalline alloys are superior, they exhibit lower saturation inductions than Fe-metalloid amorphous alloys, mainly because of the lower Fe content to attain amorphization and because of the addition of Nb and Cu (or other elements to control the nucleation and growth kinetics). In order to remedy this problem, another class of Fe-based nanocrystalline alloys. These "Nanoperm" alloys are based on the Fe-Zr-B system; they contain larger concentrations of Fe (83-89 at.%) compared to the Finemet alloys (~ 74 at.% Fe) and have higher values of saturation induction (~ 1.6-1.7 T). The Nanoperm nc alloys have very low energy losses at power frequencies (60 Hz), making them potentially interesting for electrical power distribution transformers. The issues of composition modification, processing, and the brittle mechanical behaviour of these nanocrystalline/amorphous alloys.

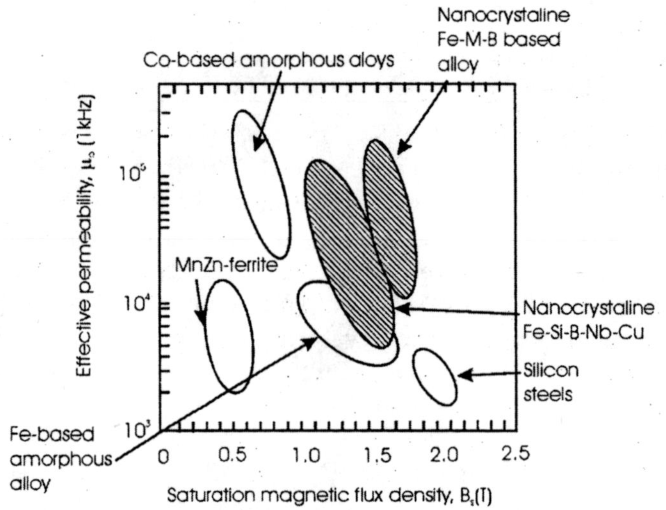


**Fig.** Effective permeability, $\mu_e$ vs. saturation magnetic flux density, $B_s$, for soft ferromagnetic materials.

The small single-domain nanocrystalline Fe particles in the amorphous matrix gives these alloys their unique magnetic behaviour, the most dramatic being the lowest energy losses of any known materials, along with very high permeabilities. These alloys can also exhibit nearly or exactly zero magnetostriction. To date, these materials have been made by crystallization of rapidly solidified amorphous ribbons. Other methods that might provide geometrically desirable products should be explored or developed. Electrodeposition is one such method that requires further work. Electrodeposited nc Fe-Ni soft magnetic alloys.

The brittle nature of these materials is a problem for scaleup and transformer manufacture. The brittleness problem must be solved by finding less brittle materials or applying the handling and processing knowledge that exists for embrittled metallic glasses.

## PERMANENT MAGNET MATERIALS

The first attempts to produce nanoscale microstructures to enhance the magnetic properties of the Nb-Fe-B permanent magnetic materials used mechanical alloying of blended elemental powders followed by heat treatment. Since the grain structure so obtained does not exhibit any crystallographic texture-and limits the energy product-special processing methods such as die-upsetting were used to provide the crystallographic anisotropy. While the coercivities of these nanocrystalline alloys are high, the remanent magnetization is decreased.

Recent approaches to increasing the magnetic induction have utilized exchange coupling in magnetically hard and soft phases. The Fe-rich compositions (e.g., $Fe_{90}Nd_7B_3$) result in a mixture of the hard $Fe_{14}Nd_2B$ phase and soft a Fe phase. The nanoscale two-phase mixtures of a hard magnetic phase and a soft magnetic phase can exhibit values of remanent magnetization, $M_r$, significantly greater than the isotropic value of 0.5 $M_s$. This "remanence enhancement" is associated with exchange coupling between the hard and soft phases, which forces the magnetization vector of the soft phase to be rotated to that of the hard phase. Two important requirements for alloys to exhibit remanence enhancement are a nanocrystalline grain size and a degree of coherence across interphase boundaries sufficient to enable adjacent phases to be exchange-coupled. The significant feature of the exchange coupling is that it allows crystallographically isotropic materials to exhibit remanence values approaching those achieved after full alignment. Such two-phase nanoscale ferromagnetic alloys have been prepared by nonequilibrium methods such as melt-spinning, mechanical alloying, and sputter deposition. Besides the high reduced remanence, the material cost is reduced by reduction in the content of the expensive hard rare earth-containing magnetic phase.

The theoretical understanding of remanence enhancement appears to be developed to a degree enabling prediction of magnet performance; however, this performance, while a significant improvement over single-phase isotropic magnets, does not reach predicted values. Work is required on optimizing the orientation relationships between the hard and soft phases and the interphase properties (coherency) between them.

Research on nanocrystalline hard magnetic alloys has received attention worldwide. While less research seems to be carried out in the world on these materials compared to the nanocrystalline soft magnetic alloys, some efforts exist in most countries. Notable programmes are those of L. Schultz and coworkers at the Institut für Festköper und Werkstofforschung (IFW) in Dresden and P.G. McCormick and coworkers at the University of Western Australia.

While the very low losses of the nc soft magnetic materials (Finemet or Nanoperm) are dependent on grain size for their properties, the hard magnetic nc alloys with remanence enhancement provide flexibility in processing, especially with powder materials. These remanence-enhanced nc hard magnetic alloys may find many applications as permanent magnet components.

## GIANT MAGNETORESISTANCE (GMR)

The phenomenon of giant magnetoresistance (GMR)-the decrease of electrical resistance of materials when exposed to a magnetic field-was first reported in a number of multilayer ferromagnetic/nonferromagnetic thin film systems. More recently, GMR was observed in equiaxed granular nanocrystalline materials. In particular, GMR systems with low saturation fields offer a wide area for application in magnetoresistive devices. GMR sensors have a higher output than conventional anisotropic magnetoresistive sensors or Hall effect sensors. They can operate at higher magnetic fields than conventional magnetoresistive sensors. In multilayer systems the antiferromagnetic alignment of the ferromagnetic layers in zero field becomes ferromagnetic as the field is applied and causes a decrease in resistance. Granular materials that show GMR consist of small ferromagnetic single-domain particles with randomly oriented magnetic axes in a nonmagnetic matrix. An external field rotates the magnetic axes of all magnetic particles. The rotation towards complete alignment of all magnetic axes again reduces the resistance in a similar way as for multilayers. The GMR in granular systems is isotropic. The explanation for the GMR is spin-dependent scattering of the conduction electrons at the ferromagnetic/nonmagnetic interfaces and, to a lesser extent, within the magnetic grains. The GMR scales inversely with the average particle diameter.

There is worldwide research on the GMR effect. U.S. programmes are reviewed by R. Shull and G.C. Hadjipanayis in the proceedings of the WTEC U.S. nanotechnologies workshop. The NIST work described by Shull has provided material with the largest GMR values for the smallest switching fields.

While the theory for GMR of spin-dependent scattering referred to above has been used as an explanation, other explanations taking into account interaction between magnetic regions. Combined theoretical and experimental studies should help to clarify the mechanism for this effect.

### Other Ferromagnetic Nanocrystalline Materials

Magnetic nanocomposite refrigerants, which have four times the magnetocaloric effects of the best low temperature magnetic refrigerant. The entropy change at a given (low) temperature for a system of magnetic spins is enhanced when the isolated spins are clustered. The nanocomposite $Gd_3Ga_{5-x}Fe_xO_{12}$ gives superior magnetocaloric effects, which increase with x up to x = 2.5 and can be extended to higher temperatures than conventional materials.

Magnetostrictive materials such as Terfonol-D ($Tb_{0.3}Dy_{0.7}Fe_2$) have been of scientific and technological interest in recent years. That nanostructured magnetostrictive materials can have improved properties, such as lower saturation fields, with reduced anisotropy and in multilayers with alternate layers of magnetostrictive and soft magnetic materials that are exchange-coupled.

## NANOCRYSTALLINE HYDROGEN STORAGE MATERIALS

Nanostructured materials offer several potential advantages for hydrogen storage materials. Rapid kinetics of absorption/desorption can be aided by refining the microstructure to the nanoscale. For example, nanoscale inclusions of $Mg_2Ni$ in Mg catalyze the decomposition of the molecular hydrogen, increasing the hydrogen absorption/desorption kinetics. Another advantage of the nanoscale microstructure is that the alloy powder does not comminute on repeated charging/discharging with hydrogen. This is not strictly a bulk material, since powder agglomerates or green compacts can be used, thus obviating the need for compaction to theoretical density.

## NANOCRYSTALLINE CORROSION-RESISTANT MATERIALS

The limited work to date on corrosion resistance of nanocrystalline materials indicates that no generalizations can be made. Superior localized corrosion resistance in HCl was observed for nanocrystalline 304 stainless steel (Fe-18%Cr-8%Ni) prepared by sputter deposition. This was attributed to the fine grain size and homogeneity of the nc material. However, the average dissolution rate of nc Ni was found to be higher than that for coarse-grained material.

# 7

# Overlayer Structures and Surface Diffraction

## CLASSIFICATION OF OVERLAYER STRUCTURES

Adsorbed species on single crystal surfaces are frequently found to exhibit long-range ordering; that is to say that the adsorbed species form a well-defined overlayer structure. Each particular structure may only exist over a limited coverage range of the adsorbate, and in some adsorbate/substrate systems a whole progression of adsorbate structures are formed as the surface coverage is gradually increased.

This section deals with the classification of such ordered structures - in most cases this involves describing the overlayer structure in terms of the underlying structure of the substrate.

There are two principal methods for specifying the structure; these are

1. Wood's notation
2. Matrix notation.

We need to make sure that we can adequately describe the structure of the substrate.

## THE CONCEPT OF THE SURFACE UNIT CELL

The primitive unit cell is the simplest periodically repeating unit which can be identified in an ordered array - the array in this instance being the ordered arrangement of surface atoms. By repeated translation of a unit cell, the whole array can be constructed.

Let us consider the clean surface structures of the low index surface planes of fcc metals...

### The fcc(100) Surface

The *fcc*(100) surface has 4-fold rotational symmetry ("square symmetry") - perhaps it should not surprise us therefore to find that the primitive unit cell for this surface is square in shape !

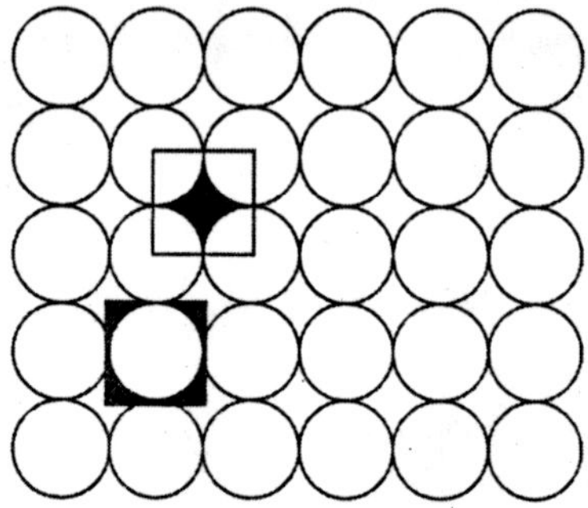

**Fig.** fcc(100) surface

Two possible choices of unit cell are highlighted - it is clear that a unit cell of this size is indeed going to be the simplest possible repeating unit for this surface. The two alternatives drawn are in fact but two of an infinte number of possibilities; they have the same shape/symmetry, size and orientation, differing only in their translational position or "origin".

Whichever we choose then it is clear that we can indeed generate the whole surface structure by repeated translation of the unit cell; for example

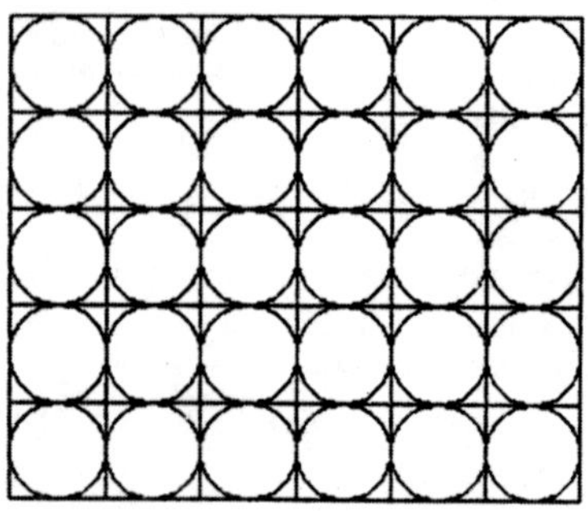

The alternative choice of unit cell which has the corners of the unit cell coincident with the atomic centres.

We now need to think how to define the unit cell shape, size and symmetry - this is best done using two vectors which have a common origin and define two sides of the unit cell.

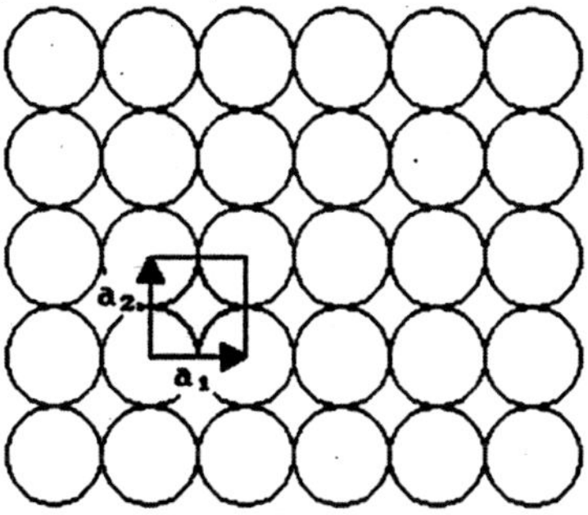


For this *fcc*(100) surface the two vectors which define the unit cell, conventionally called $a_1$ & $a_2$, are:

- The same length i.e. $|a_1| = |a_2|$
- Mutually perpendicular

By convention, one also selects the vectors such that you go anticlockwise from $a_1$ to get to $a_2$.

[Note: the length of the vectors $a_1$ & $a_2$ is related to the bulk unit cell parameter, *a*, by $|a_1| = |a_2| = a / \sqrt{2}$]

**The fcc(110) Surface**

In the case of the *fcc*(110) surface, which has 2-fold rotational symmetry, the unit cell is rectangular

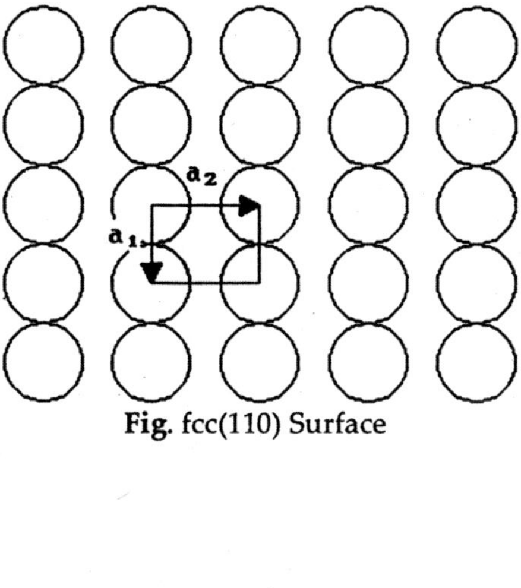


**Fig.** fcc(110) Surface

By convention, $|a_2| > |a_1|$ - if we also recall the convention that one goes anticlockwise to get from $a_1$ to $a_2$, then this leads to the choice of vectors shown.

**The fcc(111) surface**

With the *fcc*(111) surface we again have a situation where the length of the two vectors are the same i.e. $|a_1| = |a_2|$. We can either keep the angle between the vectors less than 90 degrees or let it be greater than 90 degrees. The normal convention is to choose the latter, i.e. the right hand cell of the two illustrated with an angle of 120 degrees between the two vectors.

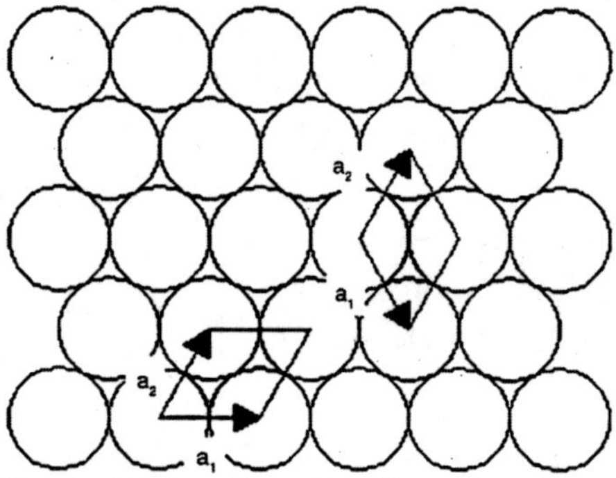

**Fig.** fcc(111) surface Overlayer Structures

If we have an ordered overlayer of adsorbed species (atoms or molecules), then we can use the same basic ideas as outlined in the previous section to define the structure.

The adsorbate unit cell is usually defined by the two vectors $b_1$ and $b_2$. To avoid ambiguities, it again helps if we stick to a set of conventions in choosing the unit cell vectors. In this case:

1. $b_2$ is again selected to be anticlockwise from $b_1$.
2. If possible, $b_1$ is chosen to be parallel to $a_1$ and $b_2$ parallel to $a_2$.

Once the unit cell vectors for substrate and adsorbate have been selected then it is a relatively simple matter to work out how to denote the structure.

## WOOD'S NOTATION

Wood's notation is the simplest and most frequently used method for describing a surface structure - it only works, however, if the two unit cells are of the same symmetry or closely-related symmetries (more specifically, the angle between $b_1$ & $b_2$ must be the same as that between $a_1$ & $a_{2)}$.

In essence, Wood's notation first involves specifying the lengths of the two overlayer vectors, $b_1$ & $b_{2,}$ in terms of $a_1$ & $a_2$ respectively - this is then written in the format:

$(|b_1|/|a_1| \times |b_2|/|a_2|)$

a ( 2 x 2 ) structure has $|b_1| = 2|a_1|$ and $|b_2| = 2|a_2|$.

The following diagram shows a ( 2 x 2 ) adsorbate overlayer on an *fcc*(100) surface in which the adsorbate is bonded terminally on-top of individual atoms of the substrate.

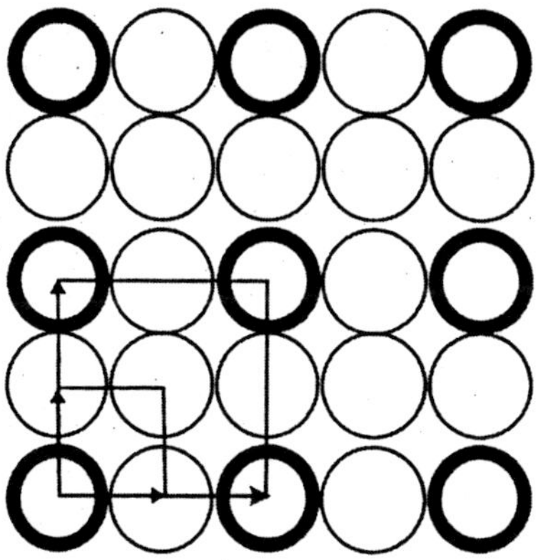

**Fig.** Substrate: fcc(100), Substrate unit cell, Adsorbate unit cell

The unit cells of the (100) substrate and the ( 2 × 2 ) overlayer are both highlighted.

The next diagram shows another ( 2 × 2 ) structure, but in this case the adsorbate species is bonded in the four-fold hollows of the substrate surface. Of course, only a very limited section of the structure can be shown here - in practice, the unit cell shown would repeat to give a complete overlayer structure extending across the substrate surface.

**Fig.** Substrate: fcc(100), Substrate unit cell, Adsorbate unit cell

The highlighted unit cells of the adsorbate and substrate are identical in size, shape and orientation to those of the previously illustrated ( 2 × 2 ) structure.

Both this and the previous structure are examples of *primitive* ( 2 × 2 ), or p( 2 × 2 ), structures. That is to say that they are indeed the simplest unit cells that may be used to describe the overlayer structure, and contain only one "repeat unit".

For the purposes of this tutorial I shall follow common practice and omit the preceding "p" - referring to such structures simply as ( 2 × 2 ) structures (in spoken language, "*two by two*" structures).

Such ( 2 × 2 ) structures are also found on other surfaces, but they may differ markedly in superficial appearance from the structure on the *fcc*(100) surface. The following diagram, for example, shows a ( 2 × 2 ) structure on a *fcc*(110) surface.

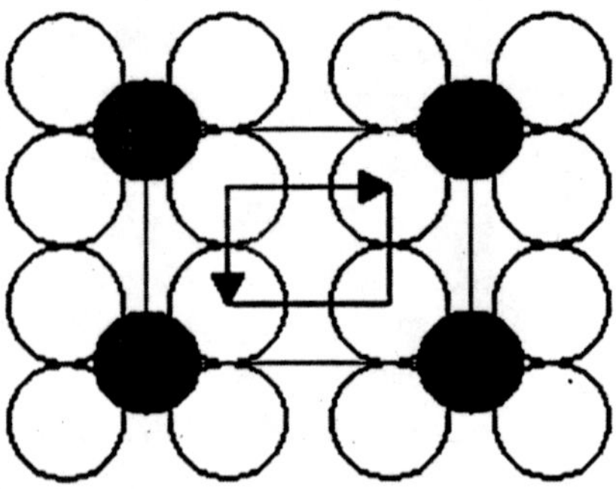

**Fig.** Substrate: fcc(110), Substrate unit cell, Adsorbate unit cell

The adsorbate unit cell is again twice as large as that of the substrate in both dimensions - it retains the same aspect ratio as the rectangular substrate unit cell (1: 1.414) and does not exhibit any rotation with respect to the substrate cell.

The following diagram shows yet another ( 2 × 2 ) structure, in this case on the fcc(111) surface...

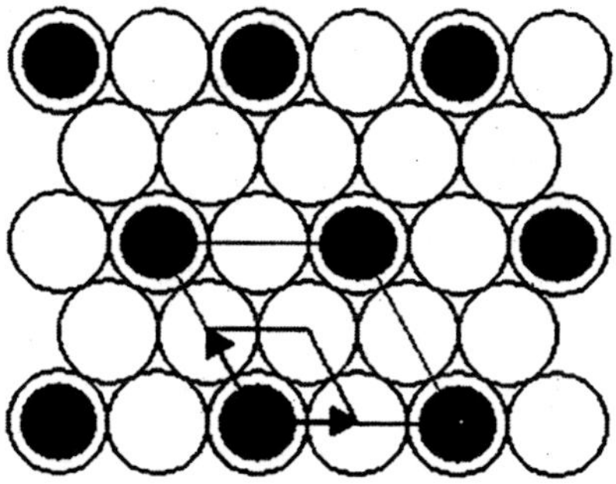

**Fig.** Substrate: fcc(111), Substrate unit cell, Adsorbate unit cell

Again, the adsorbate unit cell is of the same symmetry as the substrate cell but is scaled up by a factor of two in its linear dimensions (and corresponds to a surface area four times as large as that of the substrate unit cell).

The next example is a surface structure which is closely related to the ( 2 × 2 ) structure: it differs in that there is an additional atom in the middle/centre of the ( 2 × 2 ) adsorbate unit cell.

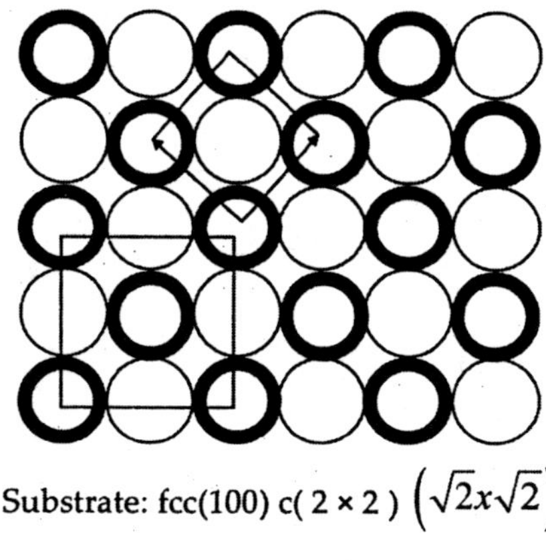

**Fig.** Substrate: fcc(100) c( 2 × 2 ) $\left(\sqrt{2}x\sqrt{2}\right)$ R45

Since the middle atom is "crystallographically equivalent" to those at the corners (i.e. it is not distinguishable by means of different coordination to the underlying substrate or any other structural feature), then this is no longer a primitive ( 2 × 2 ) structure.

Instead it may be classified in one of two ways:

1. As a *centred* ( 2 × 2 ) structure i.e. c( 2 × 2 )
2. As a " $\sqrt{2}$ (x $\sqrt{2}$ )R45 " structure, where we are specifying the true primitive unit cell.

In using the latter Wood's notation we are stating that the adsorbate unit cell is a factor of $\sqrt{2}$ larger than the substrate unit cell in both directions and is also rotated by 45 degrees with respect to the substrate unit cell.

Note: If the "central" atom is not completely crystallographically equivalent, then the structure formally remains a p(2x2) unit cell but now has a basis of two adsorbate atoms per unit cell.

In some instances it is possible to use a centred unit cell description for a structure for which the primitive unit cell cannot be described using Wood's notation - for example, the c( 2 x 2 ) structure on the *fcc*(110) surface shown below.

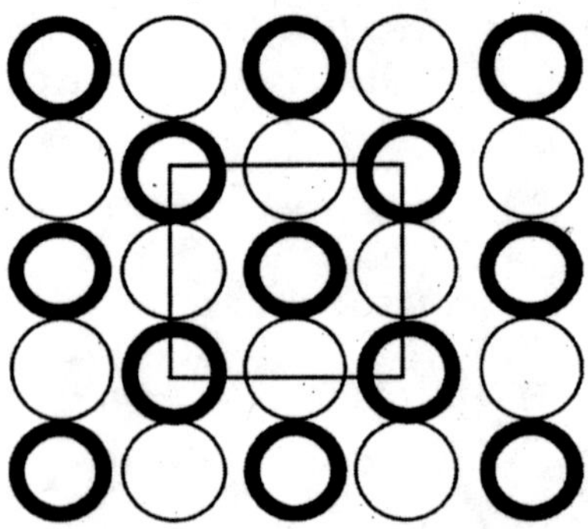

**Fig.** Substrate: fcc(110) c( 2 × 2 )

As a final example, the next diagram illustrates a commonly-observed structure on *fcc*(111) surfaces which can be readily described using Wood's notation.

**Fig.** Substrate: fcc(111) $\left(\sqrt{3}\times\sqrt{3}\right)$ R30

## MATRIX NOTATION

This is a much more general system of describing surface structures which can be applied to all ordered overlayers: quite simply it relates the vectors $b_1$ & $b_2$ to the substrate vectors $a_1$ & $a_2$ using a simple matrix i.e.

$$\begin{pmatrix} b_1 \\ b_2 \end{pmatrix} = \begin{pmatrix} * & * \\ * & * \end{pmatrix} \begin{pmatrix} a_1 \\ a_2 \end{pmatrix}$$

Matrix Notation - remember $a_1$, $a_2$, $b_1$ and $b_2$ are vectors.

To illustrate the use of matrix notation we shall now consider two surface structures with which we are already familiar...

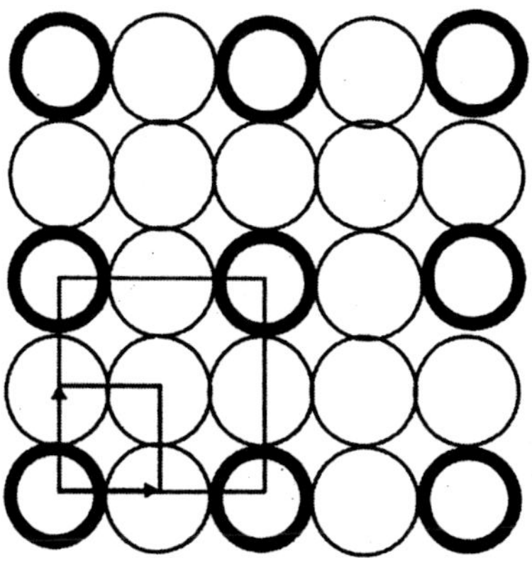

**Fig.** Substrate: fcc (100) (2 × 2) overlayer

For the (2 × 2) structure we have:

$$b_1 = 2.a_1 + 0.a_2$$
$$b_2 = 0.a_1 + 2.a_2$$
$$\Rightarrow M = \begin{bmatrix} 2 & 0 \\ 0 & 2 \end{bmatrix}$$

By contrast, for the c(2 x 2) structure:

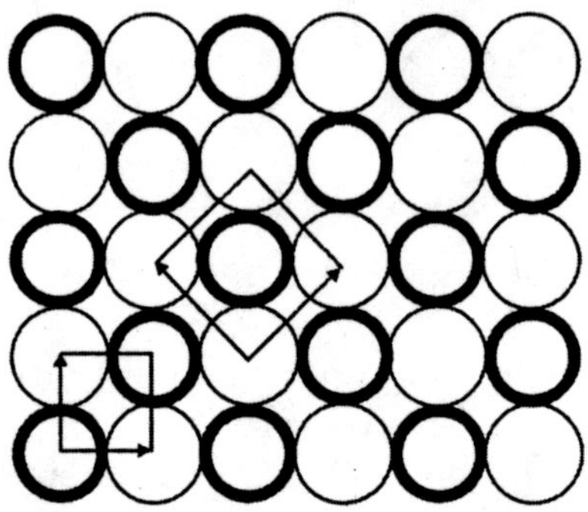

**Fig.** Substrate: fcc (100) c(2 × 2) overlayer

we have

$$b_1 = 1.a_1 + 1.a_2$$
$$b_2 = -1.a_1 + 1.a_2$$
$$\Rightarrow M = \begin{bmatrix} 1 & 1 \\ -1 & 1 \end{bmatrix}$$

Ordered surface structures may be described by defining the adsorbate unit cell in terms of that of the underlying substrate using:

**Wood's Notation**

In which the lengths of $b_1$ and $b_2$ are given as simple multiples of $a_1$ and $a_2$ respectively, and this is followed by the angle of rotation of $b_1$ from $a_1$ (if this is non-zero).

**Matrix Notation**

In which $b_1$ and $b_2$ are independently defined as linear combinations of $a_1$ and $a_2$ and these relationships are expressed in a matrix format.

## LOW ENERGY ELECTRON DIFFRACTION (LEED)

LEED is the principal technique for the determination of surface structures: it may be used in one of two ways -

1. *Qualitatively:* where the diffraction pattern is recorded and analysis of the spot positions yields information on the size, symmetry and rotational alignment of the adsorbate unit cell with respect to the substrate unit cell.
2. *Quantitatively:* where the intensities of the various diffracted beams are recorded as a function of the incident electron beam energy to generate so-called I-V curves which, by comparison with theoretical curves, may provide accurate information on atomic positions.

In this section, we will only consider the qualitative application of this experimental technique.

### Experimental Details

The LEED experiment uses a beam of electrons of a well-defined low energy (typically in the range 20 - 200 eV) incident normally on the sample. The sample itself must be a single crystal with a well-ordered surface structure in order to generate a back-scattered electron diffraction pattern. A typical experimental set-up is shown below.

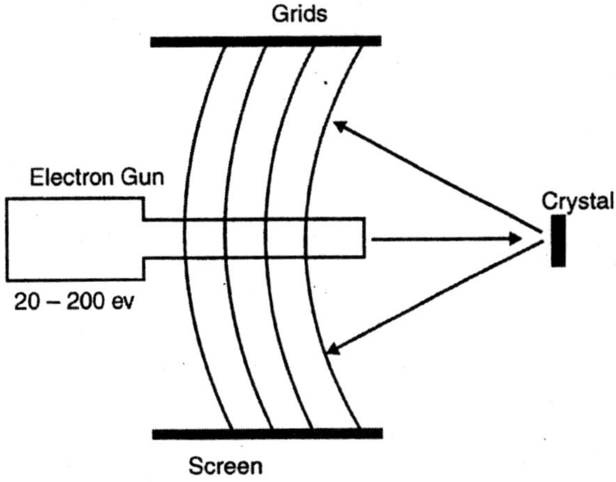

Only the elastically-scattered electrons contribute to the diffraction pattern; the lower energy (secondary) electrons are removed by energy-filtering grids placed in front of the fluorescent screen that is employed to display the pattern.

## BASIC THEORY OF LEED

By the principles of wave-particle duality, the beam of electrons may be equally regarded as a succession of electron waves incident normally on the sample. These waves will be scattered by regions of high localised electron density, i.e. the surface atoms, which can therefore be considered to act as point scatterers.

The wavelength of the electrons is given be the de Broglie relation:

Wavelength, $\lambda = h / p$ (where $p$ - electron momentum)

Now,

$p = m.\mathrm{v} = (2mE_{k)}^{1/2} = (2m.e.V)^{1/2}$

Where $m$ - mass of electron [kg]

v - velocity [m $s^{-1}$]

$E_k$ - kinetic energy

$e$ - electronic charge

$V$ - acceleration voltage (= energy in eV)

=> Wavelength, $\lambda = h / (2m.e.V)^{1/2}$

($h = 6.62 \times 10^{-34}$ J s, $e = 1.60 \times 10^{-19}$ C, $m_e = 9.11 \times 10^{-31}$ kg).

From the above examples the range of wavelengths of electrons employed in LEED experiments is seen to be comparable with atomic spacings, which is the necessary condition for diffraction effects associated with atomic structure to be observed.

Consider, first, a one dimensional (1-D) chain of atoms (with atomic separation $a$) with the electron beam incident at right angles to the chain. This is the simplest possible model for the scattering of electrons by the atoms in the topmost layer of a solid; in which case the diagram below would be representing the solid in cross-section with the electron beam incident normal to the surface from the vacuum above.

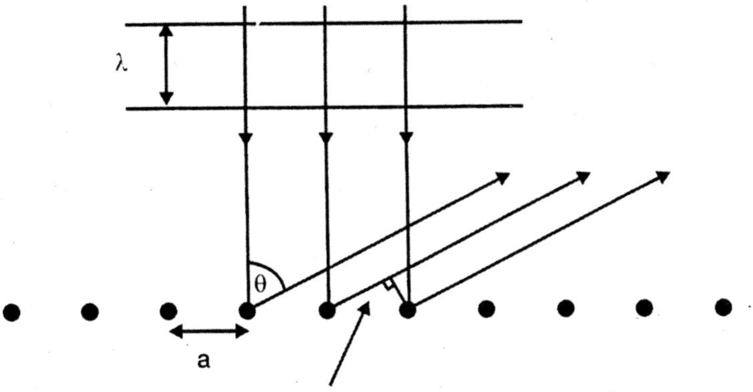


If you consider the backscattering of a wavefront from two adjacent atoms at a well-defined angle, θ, to the surface normal then it is clear that there is a "path difference" (*d*) in the distance the radiation has to travel from the scattering centres to a distant detector (which is effectively at infinity) - this path difference is best illustrated by considering two "ray paths" such as the right-hand pair of green traces.

The size of this path difference is *a sin* θ and this must be equal to an integral number of wavelengths for constructive interference to occur when the scattered beams eventually meet and interfere at the detector i.e.

$d = a$ sin q $= n$.l

where: l – wavelength

$n$ – integer (..–1, 0, 1, 2,..)

For two isolated scattering centres the diffracted intensity varies slowly between zero (complete destructive interference; $d = (n + ½)\lambda$) and its maximum value (complete constructive interference; $d = n.\lambda$) – with a large periodic array of scatterers, however, the diffracted intensity is only significant when the "Bragg condition"

$a \sin\theta = n.\lambda$

is satisfied exactly. The diagram below shows a typical intensity profile for this case.

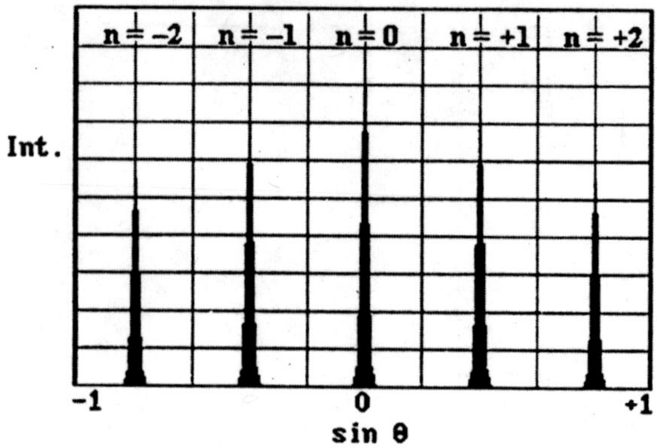


There are a number of points worth noting from this simple 1-D model

1. The pattern is symmetric about θ = 0 (or sin θ = 0)
2. Sin θ is proportional to $1 / V^{1/2}$ (since λ is proportional to $1 / V^{1/2}$)
3. Sin θ is inversely proportional to the lattice parameter, *a*

The aforementioned points are in fact much more general - all surface diffraction patterns show a symmetry reflecting that of the surface structure, are centrally symmetric, and of a scale showing an inverse relationship to both the square root of the electron energy and the size of the surface unit cell.

As an example we can look at the LEED pattern from an *fcc*(110) surface. In the diagram below the surface atomic structure is shown on the left in plan view, as if you are viewing it from the position of the electron gun in the LEED experiment (albeit greatly magnified). The primary electron beam would then be incident normally on this surface as if fired from your current viewpoint and the diffracted beams would be scattered from the surface back towards yourself. The diffraction pattern on the right illustrates how these diffracted beams would impact upon the fluorescent screen.

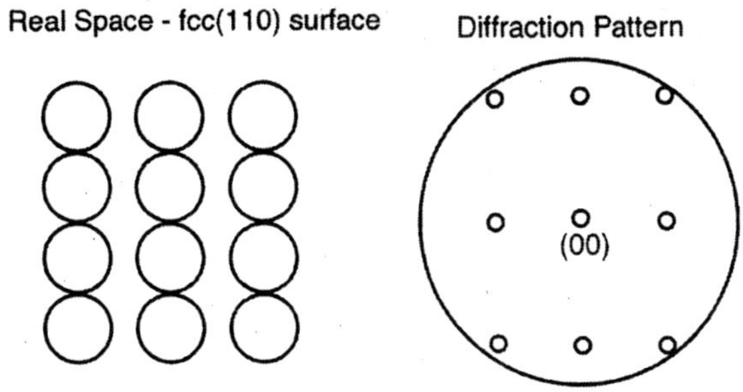


The pattern shows the same rectangular symmetry as the substrate surface but is "stretched" in the opposite sense to the real space structure due to the reciprocal dependence upon the lattice parameter. The pattern is also centrosymmetric about the (00) beam - this is the central spot in the diffraction pattern corresponding to the beam that is diffracted back exactly normal to the surface (i.e. the $n = 0$ case in our 1-D model).

The above illustration of the diffraction pattern shows only the "first-order" beams i.e. it is representative of the diffraction pattern visible at low energies when only for $n = 1$ is the angle of diffraction, q, sufficiently small for the diffracted beam to be incident on the display screen.

By contrast, the diagram below shows the diffraction pattern that might be expected if the energy of the incident electrons is doubled - some of the second order spots are now visible and the pattern as a whole has apparently contracted in towards the central (00) spot.

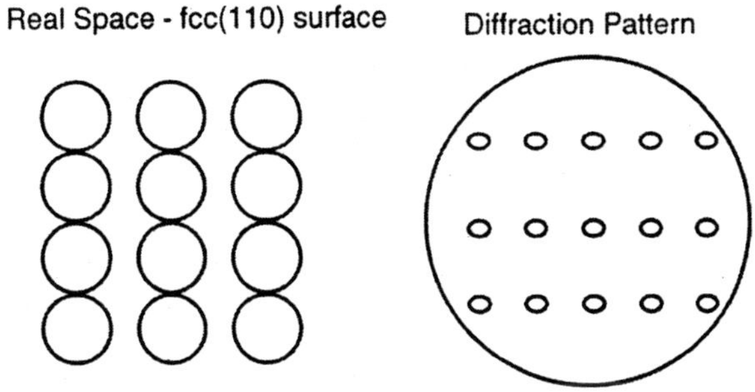

In the case of such simple LEED patterns, it is possible to explain the diffraction pattern in terms of scattering from rows of atoms on the surface. For example, the rows of atoms running vertically on the screen would give rise to a set of diffracted beams in the horizontal plane, perpendicular to the rows, thus leading to the row of spots running in a line horizontally across the diffraction pattern through the (00) spot. The further the rows are apart, then the closer in are the diffracted beams to the central (00) beam. This is, however, a far from satisfactory method of explaining LEED patterns from surfaces.

A much better method of looking at LEED diffraction patterns involves using the concept of reciprocal space: more specifically, it can be readily shown that -

**The Observed LEED Pattern is a (scaled) Representation of The Reciprocal Net of The Pseudo-2D Surface Structure "**

The *reciprocal net* is determined by (defined by) the reciprocal vectors:

$a_1^*$ and $a_2^*$ (for the substrate) and $b_1^*$ and $b_2^*$ (for the adsorbate)

Initially we will consider just the substrate. The reciprocal vectors are related to the real space unit cell vectors by the scalar product relations:

$a_1 . a_2^* = a_1^* . a_2 = 0$ and $a_1 . a_1^* = a_2 . a_2^* = 1$

For those not too keen on vector algebra these mean that:

- $a_1$ is perpendicular to $a_2^*$, and $a_2$ is perpendicular to $a_1^*$
- There is an inverse relationship between the lengths of $a_1$ and $a_1^*$ (and $a_2$ and $a_2^*$) of the form:
- $|a_1| = 1 / (|a_1^*| \cos A)$, where A is the angle between the vectors $a_1$ and $a_1^*$.

*Note:* when A = 0 degrees (cos A = 1) this simplifies to a simple reciprocal relationship between the lengths $a_1$ and $a_1^*$.

Exactly analogous relations hold for the real space and reciprocal vectors of the adsorbate overlayer structure: $b_1$, $b_1^*$, $b_2$ and $b_2^*$.

To a first approximation, the LEED pattern for a given surface structure may be obtained by superimposing the reciprocal net of the adsorbate overlayer (generated from $b_1^*$ and $b_2^*$) on the reciprocal net of the substrate (generated from $a_1^*$ and $a_2^*$)

*Example:*

Let us now look at an example - the diagram below shows an *fcc*(100) surface (again in plan view) and its corresponding diffraction pattern (i.e. the reciprocal net).

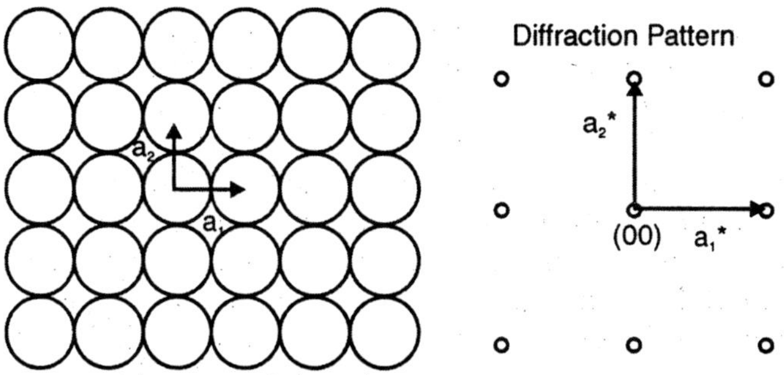


We can demonstrate how these reciprocal vectors can be determined by working through the problem in a parallel fashion for the two vectors:

$a_1^*$ must be perpendicular to $a_2$ $a_2^*$ must be perpendicular to $a_1$

⇒ $a_1^*$ is parallel to $a_1$ $a_2^*$ is parallel to $a_2$

⇒ The angle, A, between $a_1$ & $a_1^*$ is zero The angle, A, between $a_2$ & $a_2^*$ is zero

⇒ Hence, $|a_1^*| = 1 / |a_1|$ Hence, $|a_2^*| = 1 / |a_2|$

⇒ If we let $|a_1| = 1$ unit, then $|a_1^*| = 1$ unit. $|a_2| = |a_1| = 1$ unit, therefore $|a_2^*| = 1$ unit.

Let us now add in an adsorbate overlayer - a primitive (2 × 2) structure with the adsorbed species shown bonded in on-top sites - and apply the same logic as just used above to determine the reciprocal vectors, $b_1^*$ and $b_2^*$, for this overlayer.

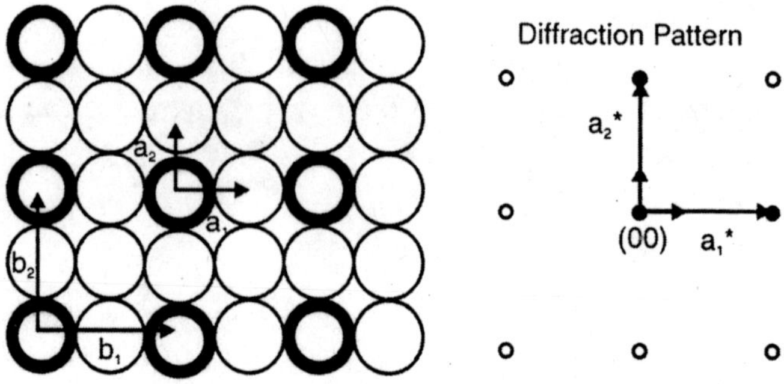


$b_1^*$ must be perpendicular to $b_2$ $b_2^*$ must be perpendicular to $b_1$

⇒ $b_1^*$ is parallel to $b_1$ $b_2^*$ is parallel to $b_2$

⇒ The angle, B, between $b_1$ & $b_1^*$ is zero The angle, B, between $b_2$ & $b_2^*$ is zero

⇒ Hence, $| b_1^* | = 1 / | b_1 |$ Hence, $| b_2^* | = 1 / | b_2 |$

⇒ $| b_1 | = 2 | a_1 | = 2$ units; Þ $| b_1^* | = ½$ unit. $| b_2 | = 2 | a_2 | = 2$ units; Þ $| b_2^* | = ½$ unit.

All we have to do now is to generate the reciprocal net for the adsorbate using $b_1^*$ and $b_2^*$.

That's all there is to it !

*Example:*

For the second example, we will look at the c(2 x 2) structure on the same *fcc*(100) surface. The diagram below shows both a real space c( 2 x 2 ) structure and the corresponding diffraction pattern:

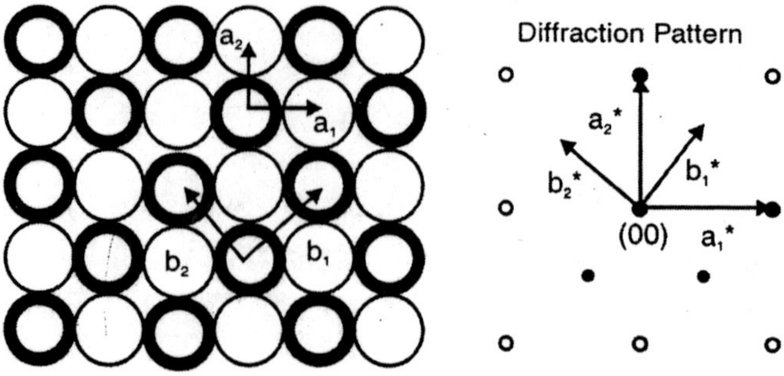

In many respects the analysis is very similar to that for the p( 2 × 2 ) structure, except that:

1. $| b_1 | = | b_2 |$ = Ö2 units; consequently $| b_1^* | = | b_2^* |$ $= 1/\sqrt{2}$ units.
2. The vectors for the adsorbate overlayer are rotated with respect to those of the substrate by 45°.

Note that the c( 2 × 2 ) diffraction pattern can also be obtained from the pattern for the primitive structure by "missing out every alternate adsorbate-derived diffraction spot". This is a common feature of diffraction patterns arising from centred structures.

## REFLECTION HIGH ENERGY ELECTRON DIFFRACTION (RHEED)

Low Energy Electron Diffraction (LEED) utilizes the inherent surface sensitivity associated with low energy electrons in order to sample the surface structure. As the primary electron energy is increased not only does the surface specificity decrease but two other effects are particularly noticeable

1. Forward scattering becomes much more important (as opposed to the backward scattering observed in LEED)
2. The scattering angle (measured from the incident beam direction) tends towards 180 degrees for back-scattering and 0 degrees for forward scattering.

In order to extract surface structural information from the diffraction of high energy electrons, therefore, the technique has to be adapted and the easiest way of doing this is to use a reflection geometry in which the electron beam is incident at a very grazing angle - it is then known as Reflection High Energy Electron Diffraction (RHEED).

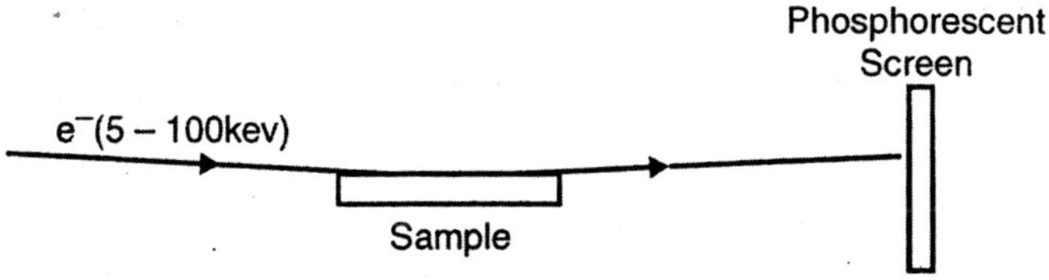

The diagram above shows the basic set-up for a RHEED experiment, with the sample viewed edge-on. In practice the display screen is usually a phosphor coating on the inside of a vacuum window (viewport) and the diffraction pattern can be viewed and recorded from the atmospheric side of the window. The small scattering angles involved are compensated for by using relatively large sample/screen distances.

The sample can be rotated about its normal axis so that the electron beam is incident along specific crystallographic directions on the surface.

In order to understand the diffraction process we need to consider how the electron beam can interact with the regular array of surface atoms in this experimental geometry. It is worth noting, however, that the use of glancing incidence ensures that, despite the high energy of the electrons, the component of the electron momentum perpendicular to the surface is small. Under these conditions an electron may travel a substantial distance through the solid (in accord with the much longer mean free path of such high energy electrons) without penetrating far into the solid. The technique, consequently, remains surface sensitive.

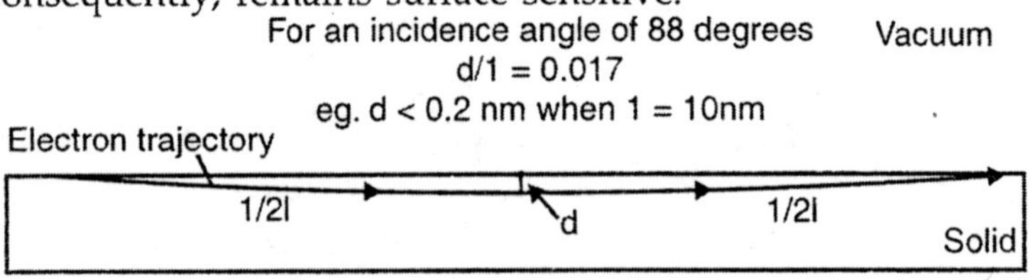


Now consider the plan view of a surface illustrated below in which we concentrate attention on just one row of atoms running in a direction perpendicular to the incident electron beam

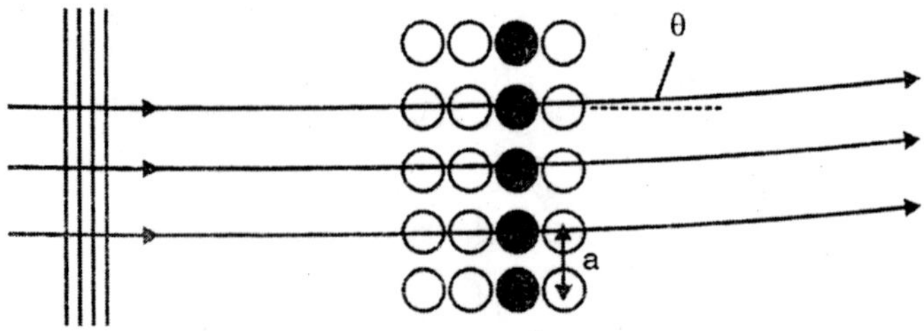

In addition to the change in momentum of the electron perpendicular to the surface, which leads to the apparent reflection, the diffraction process may also lead to a change in momentum parallel to the surface, which leads to the deflection by an angle q when looked at in plan view. Constructive interference occurs when the path difference between adjacent scattered "rays" (*a* sin q) is an integral number of wavelengths (i.e. the same basic condition as for LEED). This gives rise to a set of diffracted beams at various angles on either side of the straight through (specularly reflected) beam.

## WHAT, IF ANY, ADVANTAGES DOES RHEED OFFER OVER LEED ?

In terms of the quality of the diffraction pattern absolutely none ! - moreover, diffraction patterns have to be observed for at least two sample alignments with respect to the incident beam in order to determine the surface unit cell.

1. The geometry of the experiment allows much better access to the sample during observation of the diffraction pattern. This is particularly important if it is desired to make observations of the surface structure during growth of a surface film by evapouration from sources located normal to the sample surface or simultaneous with other measurements.
2. Experiments have shown that it is possible to monitor the atomic layer-by-atomic layer growth of epitaxial films by monitoring oscillations in the intensity of the diffracted beams in the RHEED pattern.

By using RHEED it is therefore possible to measure, and hence also to control, atomic layer growth rates in Molecular Beam Epitaxy (MBE) growth of electronic device structures - this is by far and away the most important application of the technique.

## EXAMPLES - SURFACE STRUCTURES

This section provides worked exercises in the classification of surface structures using both Wood's notation and matrix notation, and in the determination of surface coverages.

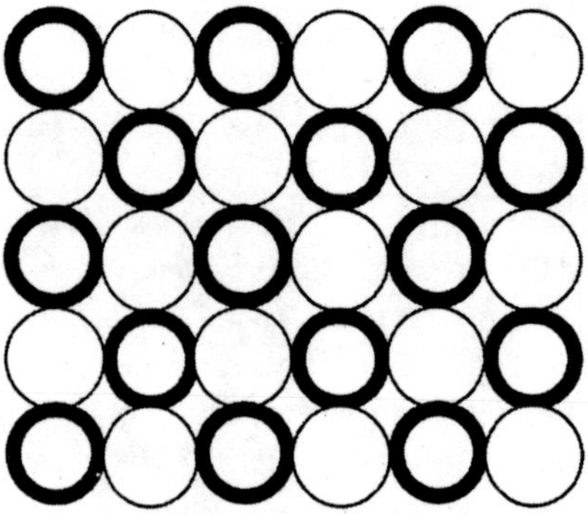

Fig. Structure 1

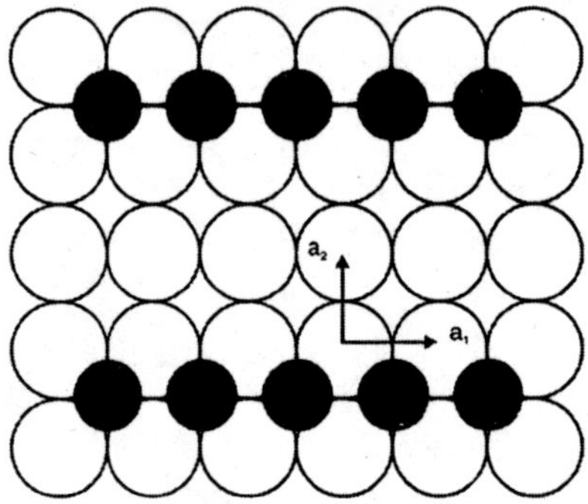


Fig. Structure 2

In this instance the structure illustrated is but one of two equivalent domains - related by the symmetry of the substrate. The second domain is shown below.

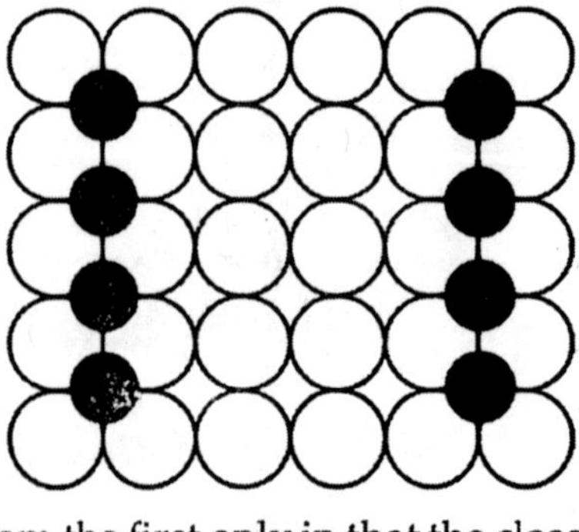

It differs from the first only in that the closely-packed rows of the adsorbate now run at 90 degrees to their direction in the first domain. Patches of this domain structure would exist on the surface with statistically-equal probability to the other domain. It obviously corresponds to the same adsorbate coverage and must possess identical properties (electronic, thermodynamic & reactivity).

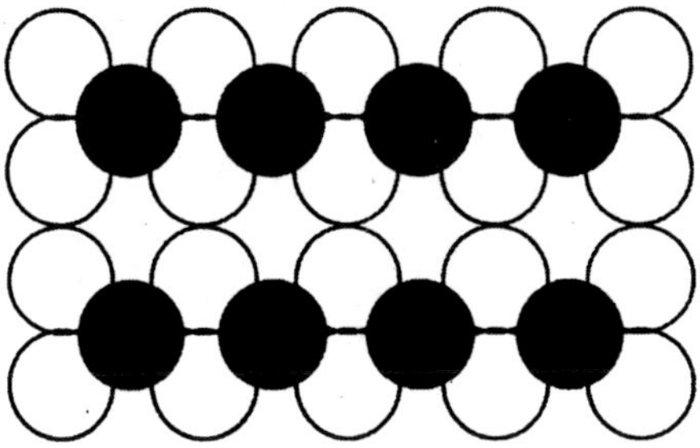

**Fig.** Structure 3

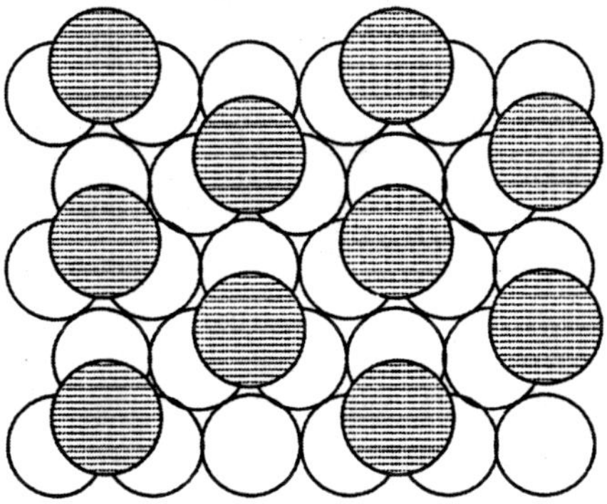

**Fig.** Structure 4

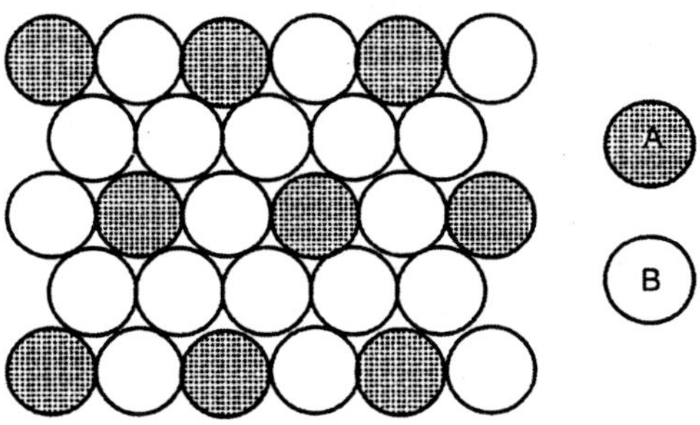


**Fig.** Structure 5

This shows an alloy surface layer on an *fcc*(111) substrate - such a layer might be formed by two metals which alloy if one is evapourated onto a (111) single crystal surface of the other.

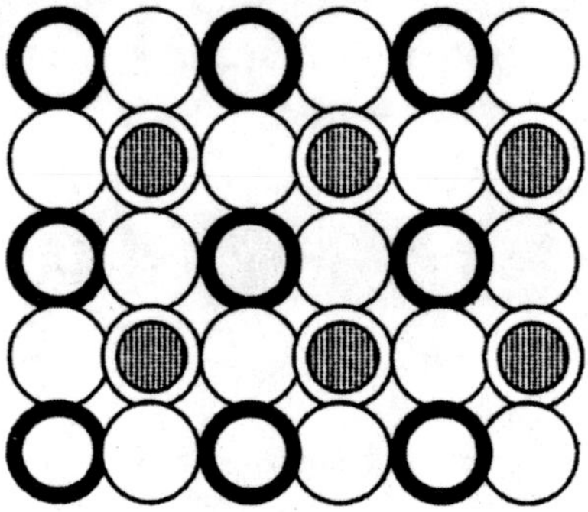

**Fig.** Structure 6

The diagram below shows a "coadsorption" structure in which adsorption of two different species has led to the formation of an ordered surface layer containing both species - such structures (albeit more complicated than the one shown) are known for a number of pairs of adsorbates on a variety of surfaces e.g. benzene / CO - Rh(111)

The driving force for the formation of such structures may be an attractive dipole-dipole force between the two different adsorbed species (this might occur if, for example, one acts as an electron donor whereas the other acts as an electron acceptor on the substrate concerned).

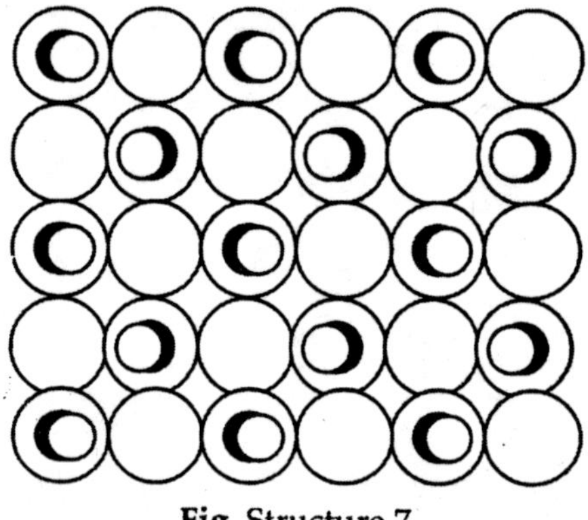

**Fig.** Structure 7

The diagram below shows a molecular adsorption structure in which a diatomic molecule is bonded terminally to substrate atoms but is inclined to the surface normal. This particular structure has not been observed but similar structures involving a periodic tilting of the molecular axis have been discovered, e.g. CO / Pd(110).

The structures of ordered adsorbate overlayers may be defined by specifying the adsorbate unit cell in terms of the ideal substrate unit cell - in many cases, such as the examples given, this can be done using the simple Wood's notation and this is the common practice. In more difficult cases it may be necessary to use matrix notation.

In all the examples studied:

1. The overlayers were simple structures in which the adsorbate unit cell dimensions corresponded to one of the atom-atom spacings in the underlying substrate structure; this is generally the case for adsorbed overlayers of gaseous molecules. More complex "coincidence" and "incommensurate" structures are commonplace when the overlayer consists of one or more atomic layers of a distinct chemical compound e.g. oxide overlayers on metals.
2. It was assumed that the underlying substrate structure was undisturbed by the adsorption of species on the surface. Whilst such distortions may generally be small, this is far from always being the case and a number of examples involving adsorbate-induced reconstruction of the topmost layer of substrate atoms have now been well documented (e.g. Ni(100) c(2 × 2)-C).

# 8

# Nanostructure

The field of nanostructure science and technology is a broad and interdisciplinary area of worldwide research and development activity that has been growing explosively in the past few years. While an understanding of the range and nature of functionalities that can be accessed through nanostructuring is just beginning to unfold, its tremendous potential for revolutionizing the ways in which materials and products are created is already clear. It is already having a significant commercial impact, and it will very certainly have a much greater impact in the future.

The World Technology Evaluation Centre (WTEC) conducted a worldwide study of the research and development status and trends in nanoparticles, nanostructured materials, and nanodevices, or more concisely, nanostructure science and technology. The study was commissioned and sponsored by a wide range of U.S. government agencies led by the National Science Foundation (NSF), which included the Air Force Office of Scientific Research, Office of Naval Research, Department of Commerce (including the National Institute of Standards and Technology and the Technology Administration), Department of Energy, National Institutes of Health, and National Aeronautics and Space Administration. Additional participating U.S. government agencies for the study were the Army Research Office, Army Research Laboratory, Defence Advanced Research Projects Agency, and the Ballistic Missile Defence Organization. The uniquely broad sponsor list for this

WTEC study mirrors the broadly based interests in and, in fact, the reality of the field of nanostructure science and technology.

The purposes of this study, which the panel determined in conjunction with its sponsors, were to assess the current status and future trends internationally in research and development in the broad and rapidly growing area of nanostructure science and technology. The study had the following four goals:

1. To provide the worldwide science and engineering community with a broadly inclusive and critical view of this field
2. To identify promising areas for future research and commercial development
3. To help stimulate development of an interdisciplinary international community of nanostructure researchers
4. To encourage and identify opportunities for international collaboration

Based on these goals, the panel formulated a number of questions, for which we sought answers during our study:

- What are the scientific drivers (new properties and phenomena, instruments, theory, and simulation methods) and advantages (applications) to be gained from control at the nanostructure level?
- What are the critical parameters to control in nanostructured material synthesis and device manufacturing?
- What are the likelihood of and the time scale for bringing these new technologies to fruition?
- What are the underlying research and development and educational concepts and directions driving nanostructure science and technology development?
- What are the expected financial dimensions of this effort over the next five to ten years, and are there national programmes in force or planned?
- Which areas of nanostructure science and technology would be most fruitful for international collaboration?

- These questions were posed in advance to the various workshop participants and hosts of our panel visits so that answers could be considered and prepared. In every case, our hosts went to great lengths and considerable efforts to prepare for our visits and to make our study both effective and comfortable.

These activities represent a rather broad base of information for the study from which the panel derived the findings and conclusions that appear in this volume. One must emphasize, however, that even though we visited many places, listened to many presentations, and read much material, there is no way that this study is, or could have been with the available resources, encyclopedic. The field of nanostructure science and technology is simply too large, too geographically dispersed, and changing too rapidly to cover exhaustively. What this volume presents are only examples, the best examples the panel could find, to describe what the field encompasses, its current breath and depth, and where it appears to be heading. The choices of the places that we visited, and even the types of visits, were made from lists and suggestions generated by all the panel members, with useful sponsor input. The final priorities were made according to where we felt the most exciting research and development activities in nanostructure science and technology were going on, overlayed with a realistic evaluation of which sites and how many of them could be logically visited in the time allotted. Unfortunately, this means that panel members on our limited schedule could not visit many interesting institutions and could not accommodate visits to entire countries that make significant contributions to nanostructure science and technology.

There are two overarching findings from this WTEC study. First, it is now abundantly clear that we are able to nanostructure materials for novel performance. This is the essential theme of this field: novel performance through nanostructuring. Nanostructuring represents the beginning of a revolutionary new age in our ability to manipulate materials for the good of humanity. The synthesis and control of materials in nanometer dimensions can access new material properties and device characteristics in unprecedented ways. Panelists had seen the tip of the iceberg or the pinnacle of the pyramid before starting this study, but only since undertaking the study do we fully appreciate just how broad the field really is and begin to understand what its exciting potential and impact may really be. It is now clear that work is rapidly expanding worldwide in exploiting the opportunities offered through nanostructuring.

The second major finding is that there is a wide range of disciplines contributing to the developments in nanostructure science and technology worldwide. Each year sees an ever increasing number of researchers from diverse disciplines enter the field and an increasing breadth of novel ideas and exciting new opportunities explode on the international nanostructure scene. The rapidly mounting level of interdisciplinary activity in nanostructuring is truly exciting. The intersections between the various disciplines are where much of the novel activity resides, and this activity is growing in importance.

These are the two basic findings that you need to carry away from this study.

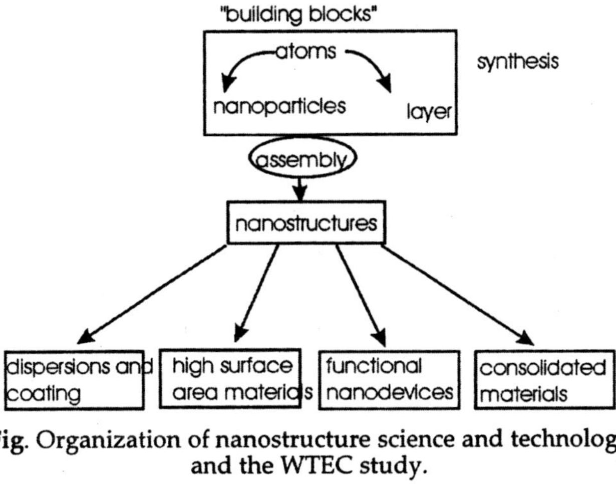


**Fig.** Organization of nanostructure science and technology and the WTEC study.

The broad field of nanostructure science and technology can be most conveniently organized according to the chart, which indicates its tremendous breadth and potential impact. The WTEC study was organized along the same lines as is this final report. The basis of the field is any type of material (metal, ceramic, polymer, semiconductor, glass, composite) created from nanoscale building blocks (clusters or nanoparticles, nanotubes, nanolayers, etc.) that are themselves synthesized from atoms and molecules. Thus, the controlled synthesis of those building blocks and their subsequent assembly into nanostructures is one fundamental theme of this field. This theme draws upon all of the materials-related disciplines from physics to chemistry to biology and to essentially all of the engineering disciplines as well. In fact, there is a very strong thread from all these disciplines running throughout the fabric of this study; the biological aspects are so pervasive that special attention.

The second and most fundamentally important theme of this field is that the nanoscale building blocks, because of their sizes below about 100 nm, impart to the nanostructures created from them new and improved properties and functionalities heretofore unavailable in conventional materials and devices. The reason for this is that materials in this size range can exhibit fundamentally new behaviour when their sizes fall below the critical length scale associated with any given property. Thus, essentially any material property can be dramatically changed and engineered through the controlled size-selective synthesis and assembly of nanoscale building blocks. Four broadly defined and overlapping application areas that cover the tremendous range of challenges and opportunities for nanostructure science and technology are dispersions and coatings, high surface area materials, functional nanodevices, and consolidated materials.

In the synthesis and assembly area we see that atoms, molecules, clusters and nanoparticles can be used as building blocks for nanostructuring. However, the useful size of these building blocks depends upon the property to be engineered, since the critical length scales for which one is designing these building blocks depends upon the particular property of interest. For multifunctional applications, more than one property and one length scale must be considered. Every property has a critical length scale, and if a nanoscale building block is made smaller than that critical length scale, the fundamental physics of that property starts to change. By altering the sizes of those building blocks, controlling their internal and surface chemistry, and controlling their assembly, it is possible to engineer properties and functionalities in unprecedented ways.

The characteristics of the building blocks, such as their size and size distribution, composition, composition variation, and morphology, must be well controlled. Also, the interfaces between the building blocks and their surroundings can be critical to performance. It is not sufficient simply to make the building blocks; one must also worry about the structure and chemistry of their surfaces and how they will interact one with another or with a matrix in which they are embedded. There is a very wide range of diverse synthesis and assembly strategies being employed in nanostructuring, all the way from fundamental biological methods for self-assembling molecules to sophisticated chemical precipitation methods to a variety of physical and chemical aerosol techniques for making clusters or nanoparticles and then dispersing them or bringing them together in consolidated forms. All of these strategies contribute in essential ways to the growth of this field. Each may have unique capabilities that will benefit a particular property, application, or process. The most generally applicable of them are likely to have significant technological impact and commercial potential.

In the area of dispersions and coatings, a wide range of new and enhanced functionalities are now becoming available by means of nanostructuring. They cover the whole set of properties that are of interest in optical, thermal, and electrical applications. This is the most mature area of nanoscale science and technology. The many current commercial applications include printing, sunscreens, photography, and pharmaceuticals. Some examples of the present technological impact of nanostructuring are thermal and optical barriers, imaging enhancement, ink-jet materials, coated abrasive slurries, and information-recording layers. From our vantage point at present, there appears to be very strong potential impact in the areas of targeted drug delivery, gene therapy, and multifunctional coatings. Nevertheless, certain central issues must be addressed if work in this area is going to continue to affect society in meaningful new ways in the coming years. Successful nanoscale dispersions require freedom from agglomeration and surface control. Process controls are required to ensure reproducibility, reliability, and scalability. There is also a need to develop process models that lead to shorter cycle times in manufacturing, if commercialization is to be truly effective.

It is of primary importance to realise that nanostructured material building blocks have inherently high surface areas unless they are consolidated. For example, a nanoparticle 5 nm in diameter has about half of its atoms on its surface. If the nanoparticles are then brought together in a lightly assembled way, this surface area is available for a variety of useful applications. In fact, there is a wide range of new applications in high capacity uses for chemical and electrical energy storage, or in sensors and other applications that take copious advantage of this feature. Already there are numerous commercial applications in porous membranes or molecular sieves, drug delivery, tailored catalysts, and absorption/desorption materials. Clearly, what is required to optimize the impact of nanostructures to be really useful to society in high surface area material applications is to create materials that combine high selectivity, high product or function yield, and high stability. Thus, the major challenges in this area are critical dimensional control and long-term thermal and chemical stability. When these problems are solved, considerable future technological potential is seen in the areas of molecule-specific sensors, large hydrocarbon or bacterial filters, energy storage, and Grätzel-type solar cells.

The area of functional nanoscale devices is largely driven by the need for ever smaller devices, which necessitate both new device and new circuit architectures. It is not very useful to make nanoscale devices if they cannot be assembled in a circuit with interconnects that are themselves nanoscale. Thus, a complete rethinking of this area is required. The major research and development effort worldwide in functional nanoscale devices is focused on the single electron transistor (SET) using a variety of nanostructuring approaches. However, there is also considerable worldwide activity on magnetic devices using giant magnetoresistance (GMR) of nanostructures with architectures of various modulation dimensionalities. In fact, it is the nanostructuring with various modulation dimensionalities that has created an expanding range of different functionalities that can be engineered into these GMR devices. There is also exciting carbon nanotube research being actively pursued in areas of high-field-emission displays and several other nanoscale electronic devices. This is an area still very early in its development, since nanotubes and their derivatives are a relatively recent discovery, but one with tremendous potential.

While there is little technological impact already present in the nanoscale device area other than GMR read heads, several potential areas of significant impact do appear on the horizon. These include terabit memory and microprocessing; single molecule DNA sizing and sequencing; biomedical sensors; low-noise, low-threshold lasers; and nanotubes for high brightness displays. Nevertheless, a major challenge looms in the efficient manipulation of these nanoscale building blocks and their eventual commercial scaleup, if any of this is really going to affect society as we know it. One shining example that indicates probable success in overcoming such obstacles in the future is the ability to now translate SET devices made by individual atomic manipulation into arrays of similarly functional devices created by the biological self-assembly of large molecular arrays. Such cross-disciplinary transfers of nanostructuring ideas and capabilities can be expected to increasingly impact the future successful implementation of nanostructure science and technology.

We have known for about a decade that the bulk behaviour of materials can be dramatically altered when constituted of, or consolidated from, nanoscale building blocks. This can significantly and favourably affect the mechanical properties, magnetic properties, and optical properties of a range of engineering materials. We already know that the hardness and strength of nanophase metals can be greatly increased by nanostructuring, for example. On the other hand, the ductility and superplastic forming capabilities of nanophase ceramics have now become possible generically, leading to new processing routes that will be more cost-effective than present methods. Nanoparticle fillers in metal, ceramic, or polymer matrices can yield a very wide range of nanocomposites with unique properties.

This is an area that in some cases is just beginning to be researched seriously, but it could have huge technological impact in the future. Nanostructuring can also uniquely create both soft and hard magnetic materials with greatly improved performances. These materials are already having technological impact in the areas of low-loss magnets, high-hardness and tough cutting tools, and nanocomposite cements. Potential technological applications with high commercial impact can be expected in the areas of superplastic forming of ceramics, ultrahigh-strength and tough structural materials, magnetic refrigerants, a wide range of nanoparticle-filled polymer nanocomposites based on elastomers, thermoplastics and thermosets, and ductile cements. Nanoparticles, nanostructured materials, and nanodevices from the point of view of biological applications and biological analogies. Current research directed toward biological synthesis and assembly is highlighted as it pertains to the building blocks of nanotechnology, and examples are presented of state-of-the-art research on the biological aspects of dispersions and coatings, high surface area materials, and functional nanostructures. A primary finding is that although biological applications of nanostructure science and technology may not be as well developed currently as non-biological ones, they nevertheless present a very promising research and development frontier that is likely to have tremendous future impact.

It is noteworthy that these funding levels have been increasing very rapidly in recent years as the number of researchers worldwide who are excited about this field have multiplied and funding agencies have responded accordingly. The various ways in which nanostructure science and technology research is funded in the countries the panel surveyed had often appeared quite different from a distance, but are actually quite similar to one another at closer view. Some countries, most notably Japan, have tended to primarily fund their nanostructure research through large national programmes, with a rather monolithic appearance from afar, centreed at national laboratories or at major national universities. On the other hand, with some exceptions, most of the nanostructure research funding in the United States and Europe tends to be based upon competition among individual research groups for smaller amounts of support. In both types of nanostructure funding schemes, however, it seems that the individual researchers actually dominate how the work proceeds. In most cases, any significant interactions among researchers occur through normal personal and professional contacts; large-scale institutionalized cooperative research efforts in this field have often not been particularly effective. A particularly impressive national funding effort in nanostructure science and technology occurs in France under the auspices of the Centre National de la Récherche Scientifique (CNRS). There, an extensive multidisciplinary network of laboratories in universities, industries, and national laboratories, funded partly by the CNRS and partly by industry, appear to interact successfully. It could be a very useful model to follow.

# 9

# The Langmuir Isotherm

## THE LANGMUIR ISOTHERM

Whenever a gas is in contact with a solid there will be an equilibrium established between the molecules in the gas phase and the corresponding adsorbed species (molecules or atoms) which are bound to the surface of the solid.

As with all chemical equilibria, the position of equilibrium will depend upon a number of factors:

1. The relative stabilities of the adsorbed and gas phase species involved
2. The temperature of the system (both the gas and surface, although these are normally the same)
3. The pressure of the gas above the surface

In general, factors (2) and (3) exert opposite effects on the concentration of adsorbed species - that is to say that the *surface coverage* may be increased by raising the gas pressure but will be reduced if the surface temperature is raised.

The Langmuir isotherm was developed by Irving Langmuir in 1916 to describe the dependence of the surface coverage of an adsorbed gas on the pressure of the gas above the surface at a fixed temperature. There are many other types of isotherm (Temkin, Freundlich...) which differ in one or more of the assumptions made in deriving the expression for the surface coverage; in particular, on how they treat the surface coverage dependence of the enthalpy of adsorption. Whilst the

Langmuir isotherm is one of the simplest, it still provides a useful insight into the pressure dependence of the extent of surface adsorption.

When considering adsorption isotherms it is conventional to adopt a definition of surface coverage (q) which defines the maximum (*saturation*) surface coverage of a particular adsorbate on a given surface always to be unity, i.e. $q_{max} = 1$.

This way of defining the surface coverage differs from that usually adopted in surface science where the more common practice is to equate q with the ratio of adsorbate species to surface substrate atoms (which leads to saturation coverages which are almost invariably less than unity).

## LANGMUIR ISOTHERM-DERIVATION FROM EQUILIBRIUM CONSIDERATIONS

We may derive the Langmuir isotherm by treating the adsorption process as we would any other equilibrium process - except in this case the equilibrium is between the gas phase molecules (M), together with vacant surface sites, and the species adsorbed on the surface. Thus, for a non-dissociative (molecular) adsorption process we consider the adsorption to be represented by the following chemical equation:

$$S - * + M_{(g)} = S - M$$

where: S – *, represents a vacant surface site

This equation we are making an inherent assumption that there are a fixed number of localised surface sites present on the surface. This is the first major assumption of the Langmuir isotherm.

We may now define an equilibrium constant (*K*) in terms of the concentrations of "reactants" and "products"

$$K = \frac{[S-M]}{[S-*][M]}$$

We may also note that:

- [S – M] is proportional to the surface coverage of adsorbed molecules, i.e. proportional to q

- [S – *] is proportional to the number of vacant sites, i.e. proportional to (1 – q)
- [M] is proportional to the pressure of gas, $P$

Hence, it is also possible to define another equilibrium constant, $b$, as given below:

$$b = \frac{\theta}{(1-\theta)P}$$

Rearrangement then gives the following expression for the surface coverage

$$\theta = \frac{bP}{1+bP}$$

which is the usual form of expressing the Langmuir Isotherm.

As with all chemical reactions, the equilibrium constant, $b$, is both temperature-dependent and related to the Gibbs free energy and hence to the enthalpy change for the process.

b is only a constant (independent of q if the enthalpy of adsorption is independent of coverage - this is the second major assumption of the Langmuir Isotherm.

## LANGMUIR ISOTHERM - DERIVATION FROM KINETIC CONSIDERATIONS

The equilibrium that may exist between gas adsorbed on a surface and molecules in the gas phase is a dynamic state, i.e. the equilibrium represents a state in which the rate of adsorption of molecules onto the surface is exactly counterbalanced by the rate of desorption of molecules back into the gas phase. It should therefore be possible to derive an isotherm for the adsorption process simply by considering and equating the rates for these two processes.

Expressions for the rate of adsorption and rate of desorption have been derived in Sections respectively: specifically,

$$R_{ads} = \frac{f(\theta)\cdot P}{\sqrt{2\pi mKT}}\exp\left(-E_a^{ads}/RT\right)$$

Equating these two rates yields an equation of the form:

$$\frac{P.f(\theta)}{f'(\theta)} = C(T)$$

where the terms $f(\theta)$ & $f'(\theta)$ contain the pre-exponential surface coverage dependence of the rates of adsorption and desorption respectively and all other factors have been taken over to the right hand side to give a temperature-dependent "constant" characteristic of this particular adsorption process, $C(T)$.

We now need to make certain simplifying assumptions... the first is one of the key assumptions of the Langmuir isotherm.

Adsorption takes place only at specific localized sites on the surface and the saturation coverage corresponds to complete occupancy of these sites.

Let us initially further restrict our consideration to a simple case of reversible molecular adsorption..

$$S - + M_{(g)} = S - M$$

where S - represents a vacant surface site and S - M the adsorption complex.

Under these circumstances it is reasonable to assume coverage dependencies for rates of the two processes of the form:

Adsorption, $f(q) = c\,(1\text{-}q)$ i.e. proportional to the fraction of sites that are unoccupied.

Desorption, $f'(q) = c'q$ i.e. proportional to the fraction of sites which are occupied by adsorbed molecules.

where $\theta$ is the fraction of sites occupied at equilibrium.

These coverage dependencies are exactly what would be predicted by noting that the forward and reverse processes are elementary reaction steps, in which case it follows from standard chemical kinetic theory that

1. The forward adsorption process will exhibit kinetics having a first order dependence on the concentration of vacant surface sites.
2. The reverse desorption process will exhibit kinetics

having a first order dependence on the concentration of adsorbed molecules.

Substitution into equation then yields

$$\frac{P\cdot(1-\theta)}{\theta} = B(T)$$

where $B(T) = (c'/c).C(T)$. After rearrangement this gives the Langmuir Isotherm expression for the surface coverage

$$\theta = \frac{bP}{1+bP}$$

where $b$ (=1/$B(T)$) is a function of temperature and contains an exponential term of the form

$$b = \exp\left[(E_a^{des} - E_a^{ads}) / R\,T\right] = \exp\left[-DH_{ads} / R\,T\right]$$

Consequently, $b$ can only be regarded as a constant with respect to coverage if the enthalpy of adsorption is itself independent of coverage - this is the second major assumption of the Langmuir Isotherm.

## VARIATION OF SURFACE COVERAGE WITH TEMPERATURE AND PRESSURE

Application of the assumptions of the Langmuir Isotherm leads to readily derivable expressions for the pressure dependence of the surface coverage in the case of a simple, reversible molecular adsorption process the expression is

$$\theta = \frac{bP}{1+bP} \quad \text{where } b = b(\text{T})$$

This is illustrated in the graph below which shows the characteristic Langmuir variation of coverage with pressure for molecular adsorption.

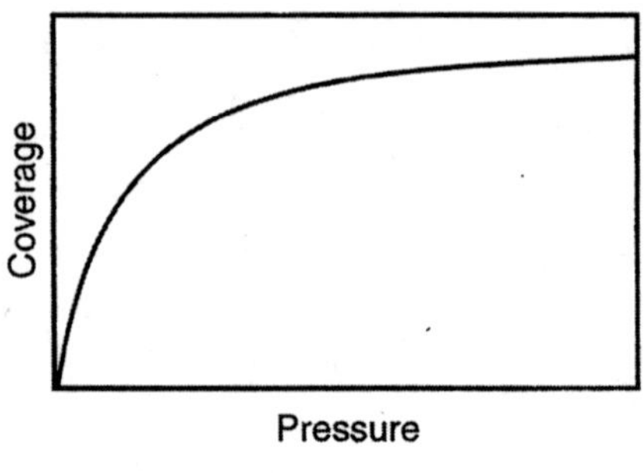

1. $q \to bP$ at low pressures
2. $q \to 1$ at high pressures

At a given pressure the extent of adsorption is determined by the value of *b*: this in turn is dependent upon both the temperature (T) and the enthalpy (heat) of adsorption. Remember that the magnitude of the adsorption enthalpy (a negative quantity itself) reflects the strength of binding of the adsorbate to the substrate.

The value of *b* is increased by

1. A reduction in the system temperature
2. An increase in the strength of adsorption

Therefore the set of curves shown below illustrates the effect of either

1. Increasing the magnitude of the adsorption enthalpy at a fixed temperature, or
2. Decreasing the temperature for a given adsorption system.

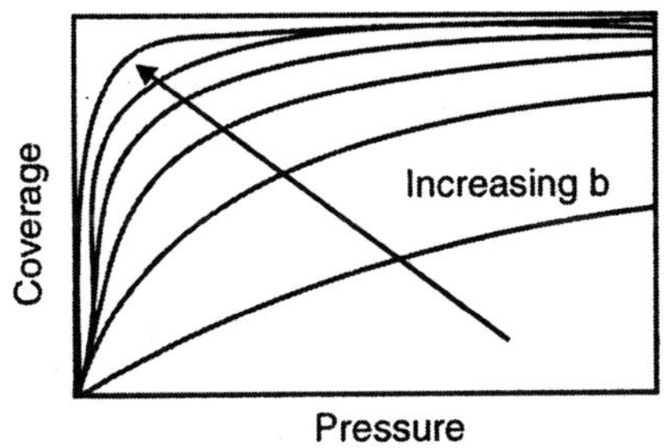


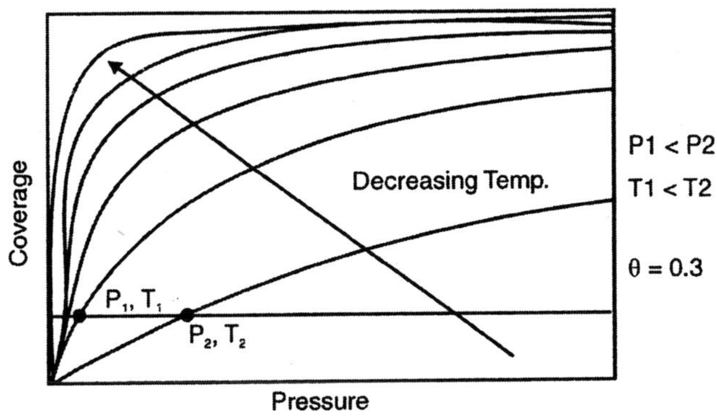

A given equilibrium surface coverage may be attainable at various combinations of pressure and temperature note that as the temperature is lowered the pressure required to achieve a particular equilibrium surface coverage decreases.

This is often used as justification for one of the main ideologies of surface chemistry; specifically, that it is possible to study technologically-important (high pressure / high temperature) surface processes within the low pressure environment of typical surface analysis systems by working at low temperatures. It must be recognised however that, at such low temperatures, kinetic restrictions that are not present at higher temperatures may become important.

## DETERMINATION OF ENTHALPIES OF ADSORPTION

It has been shown in previous sections how the value of *b* is dependent upon the enthalpy of adsorption. It has also just been demonstrated how the value of *b* influences the pressure/temperature (P-T) dependence of the surface coverage.

The implication of this is that it must be possible to determine the enthalpy of adsorption for a particular adsorbate/substrate system by studying the P-T dependence of the surface coverage.

Various methods based upon this idea have been developed for the determination of adsorption enthalpies - one method is outlined below:

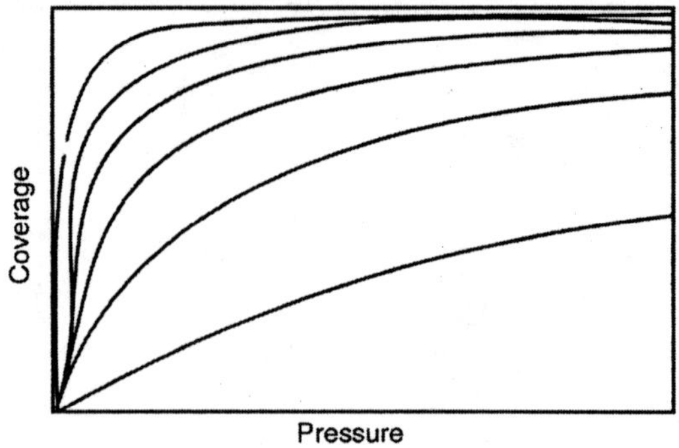

*Step 1:* Involves determination of a number of adsorption isotherms (where a single isotherm is a coverage / pressure curve at a fixed temperature).

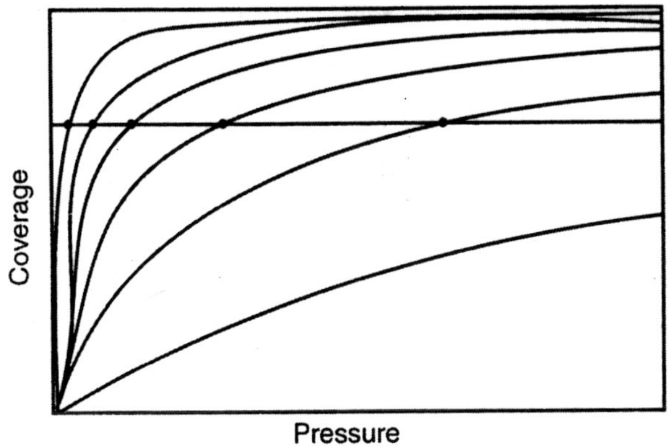


*Step 2:* It is then possible to read off a number of pairs of values of pressure and temperature which yield the same surface coverage may then be applied to this set of (P-T) data

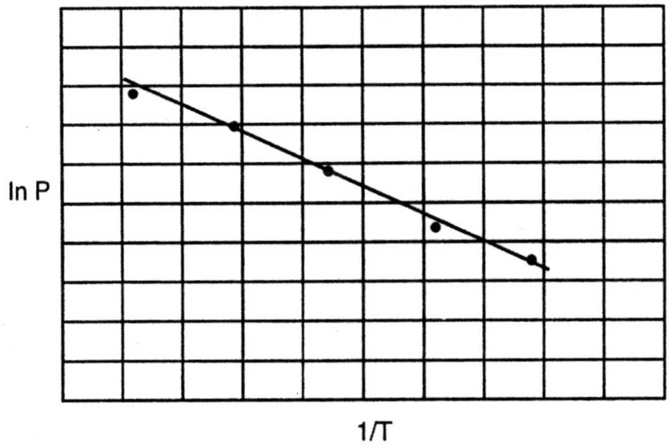


*Step 3:* The Clausius-Clapeyron equation

and a plot of (ln P) v's (1/T) should give a straight line, the slope of which yields the adsorption enthalpy.

- The value obtained for the adsorption enthalpy is that pertaining at the surface coverage for which the P-T data was obtained, but steps 2 & 3 may be repeated for different surface coverages enabling the adsorption enthalpy to be determined over the whole range of coverages.

- This method is applicable only when the adsorption process is thermodynamically reversible.

## APPLICATIONS: KINETICS OF CATALYTIC REACTIONS

It is possible to predict how the kinetics of certain heterogeneously-catalysed reactions might vary with the partial pressures of the reactant gases above the catalyst surface by using the Langmuir isotherm expression for equilibrium surface coverages.

### Unimolecular Decomposition

Consider the surface decomposition of a molecule A, i.e. the process

$$A_{(g)} \leftrightarrow A_{(ads)} \rightarrow \text{Products}$$

Let us assume that:

1. The decomposition reaction occurs uniformly across the surface sites at which molecule A may be adsorbed and is not restricted to a limited number of special sites.
2. The products are very weakly bound to the surface and, once formed, are rapidly desorbed.
3. The rate determining step (rds) is the surface decomposition step.

Under these circumstances, the molecules of A adsorbed on the surface are in equilibrium with those in the gas phase and we may predict the surface concentration of A from the Langmuir isotherm i.e.

$$\theta = b.P / (1 + b.P)$$

The rate of the surface decomposition (and hence of the reaction) is given by an expression of the form

$$\text{Rate} = k\,\theta$$

Substituting for the coverage, q, gives us the required expression for the rate in terms of the pressure of gas above the surface

Rate = $k\, b\, P\, /\, (1 + b\, P)$

It is useful to consider two extreme limits:

*Limit 1*

: $b.P << 1$; then

$(1 + b.P) \sim 1$ and Rate $\sim k.b.P$

i.e. a first order reaction (with respect to the partial pressure of A) with an apparent first order rate constant, $k' = k.b$ .

This is the low pressure (or weak binding i.e. small $b$) limit: under these conditions the steady state surface coverage, θ, of the reactant molecule is very small.

*Limit 2*

: $b.P >> 1$; then

$(1 + b.P) \sim b.P$ and Rate $\sim k$

i.e. a zero order reaction (with respect to the partial pressure of A).

This is the high pressure (or strong binding i.e. large $b$) limit: under these conditions the steady state surface coverage, θ, of the reactant molecule is almost unity.

In fact, the rate shows the same pressure variation as does the surface coverage (hardly surprising since it is directly proportional to q)

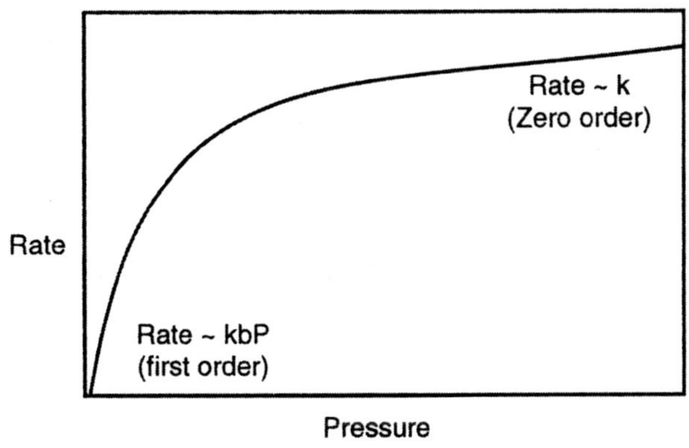

**Bimolecular Reaction (between Molecular Adsorbates)**

Consider a Langmuir-Hinshelwood reaction of the following type:

$$A_{(g)} = A_{(ads)}$$

$$B_{(g)} = B_{(ads)}$$

$$A_{(ads)} + B_{(ads)} \xrightarrow{rds} AB_{(ads)} \xrightarrow{fast} AB_{(g)}$$

We will further assume, as noted in the above scheme, that the surface reaction between the two adsorbed species is the rate determining step.

If the two adsorbed molecules are mobile on the surface and freely intermix then the rate of the reaction will be given by the following rate expression for the bimolecular surface combination step

$$\text{Rate} = k\,\theta_A\,\theta_B$$

For a single molecular adsorbate the surface coverage (as given by the Langmuir isotherm) is:

$$\theta = b.P\,/\,(1 + b.P)$$

Where two molecules (A & B) are competing for the same adsorption sites then the relevant expressions are:

$$\theta_A = \frac{b_A P_A}{1 + b_A P_A + b_B P_B} \text{ and}$$

$$\theta_B = \frac{b_B P_B}{1 + b_A P_A + b_B P_B}$$

Substituting these into the rate expression gives:

$$Rate = k\theta_A\theta_B = \frac{kb_A P_A \cdot b_B P_B}{(1 + b_A P_A + b_B P_B)^2}$$

Once again, it is interesting to look at several extreme limits

*Limit 1*

: $b_A\,P_A \ll 1$ & $b_B\,P_B \ll 1$

In this limit $q_A$ & $q_B$ are both very low, and

Rate $\rightarrow k\, b_A\, P_A\, b_B\, P_B = k'\, P_A\, P_B$

i.e. *first order* in both reactants

*Limit 2*

: $b_A\, P_A << 1 << b_B\, P_B$

In this limit $q_A \rightarrow 0$, $q_B \rightarrow 1$, and

Rate $\rightarrow k\, b_A\, P_A\, /\, (b_B\, P_{B)} = k'\, P_A\, /\, P_B$

i.e. *first order* in A, but *negative first order* in B

From this example it is readily seen that, depending upon the partial pressure and binding strength of the reactants, a given model for the reaction scheme can give rise to a variety of apparent kinetics: this highlights the dangers inherent in the reverse process - namely trying to use kinetic data to obtain information about the reaction mechanism.

**CO Oxidation Reaction**

On precious metal surfaces (e.g. Pt) the CO oxidation reaction is generally believed to by a Langmuir-Hinshelwood mechansim of the following type:

$CO_{(g)} = CO_{(ads)}$

$O_{2\,(g)} = 2\,O_{(ads)}$

$CO_{(ads)} + O_{(ads)} \xrightarrow{rds} CO_{2\,(ads)} \xrightarrow{fast} CO_{2\,(g)}$

As $CO_2$ is comparatively weakly-bound to the surface, the desorption of this product molecule is relatively fast and in many circumstances it is the surface reaction between the two adsorbed species that is the rate determining step.

If the two adsorbed molecules are assumed to be mobile on the surface and freely intermix then the rate of the reaction will be given by the following rate expression for the bimolecular surface combination step

Rate = $k\theta_{CO}\theta_O$

Where two such species (one of which is molecularly adsorbed, and the other dissociatively adsorbed) are competing for the same adsorption sites then the relevant expressions are:

$$\theta_{CO} = \frac{b_{CO}P_{CO}}{\left(1+\sqrt{b_O P_{O_2}}+b_{CO}P_{CO}\right)}$$

and

$$\theta_O = \frac{\sqrt{b_o P_c}}{\left(1+\sqrt{b_o P_{o_2}}+b_{CO}P_{CO}\right)}$$

Substituting these into the rate expression gives:

$$\text{Rate} = K\theta_{CO}\theta_0 = \frac{kb_{CO}P_{CO}\sqrt{b_O P_{O_2}}}{\left(1+\sqrt{b_O P_{O_2}}+b_{CO}P_{CO}\right)^2}$$

Once again, it is interesting to look at certain limits. If the CO is much more strongly bound to the surface such that $b_{CO}P_{CO} >> (1 + \sqrt{(b_O P_{O2}})$ and hence that $(1 + \sqrt{(b_O P_{O2})} + b_{CO}P_{CO})$ $\approx b_{CO}P_{CO}$ then the rate equation simplifies to give

$$\text{Rate} \approx k\sqrt{\frac{b_O P_{O_2}}{b_{CO}}} = k'\frac{P_{O_2}^{1/2}}{P_{CO}}$$

In this limit the kinetics are half-order with respect to the gas phase pressure of molecular oxygen, but negative order with respect to the CO partial pressure, i.e. CO acts as a poison (despite being a reactant) and increasing its pressure slows down the reaction. This is because the CO is so strongly bound to the surface that it blocks oxygen adsorbing, and without sufficient oxygen atoms on the surface the rate of reaction is reduced.

## ULTRA HIGH VACUUM (UHV)

Vacuum technology has advanced considerably over the last 25 years and very low pressures are now routinely obtainable.

Firstly, we need to remind ourselves of the units of pressure:

- The SI unit of pressure is the Pascal (1 Pa = 1 N $m^{-2}$)
- Normal atmospheric pressure (1 atm.) is 101325 Pa or 1013 mbar (1 bar = $10^5$ Pa).
- An older unit of pressure is the Torr (1 Torr = 1 mmHg). One atmosphere is ca. 760 Torr (i.e. 1 Torr = 133.3 Pa).

Whilst the mbar is often used as a unit of pressure for describing the level of vacuum, the most commonly employed unit is still the Torr. (The SI unit, the Pa, is almost never used !)

Classification of the degree of vacuum is hardly an exact science - it very much depends upon who you are talking to - but as a rough guideline:

| | |
|---|---|
| Rough (low) vacuum: | $1 - 10^{-3}$ Torr |
| Medium vacuum: | $10^{-3} - 10^{-5}$ Torr |
| High vacuum (HV): | $10^{-6} - 10^{-8}$ Torr |
| Ultrahigh vacuum (UHV): | $< 10^{-9}$ Torr |

Virtually all surface studies are carried out under UHV conditions - the question is why ?.

Ultra high vacuum is required for most surface science experiments for two principal reasons:

1. To enable atomically clean surfaces to be prepared for study, and such surfaces to be maintained in a contamination-free state for the duration of the experiment.
2. To permit the use of low energy electron and ion-based experimental techniques without undue interference from gas phase scattering.

To put these points in context we shall now look at the variation of various parameters with pressure

**Gas Density**

The gas density is easily estimated from the ideal gas law:

$n = (N / V) = P / (k.T)$ [molecules $m^{-3}$]

where: $P$ – pressure [N m$^{-2}$]

$k$ – Boltzmann constant (= 1.38 × 10$^{-23}$ J K$^{-1}$)

$T$ – temperature [K]

**Mean Free Path of Particles in the Gas Phase**

The average distance that a particle (atom, electron, molecule..) travels in the gas phase between collisions can be determined from a simple hard-sphere collision model this quantity is known as the *mean free path* of the particle, and for neutral molecules is given by the equation:

$$\lambda = \frac{kT}{1.414 P\sigma}$$

where: $P$ – pressure [N m$^{-2}$]

$k$ – Boltzmann constant (= 1.38 × 10$^{-23}$ J K$^{-1}$)

$T$ – temperature [K]

s – collision cross section [m$^2$]

**Incident Molecular Flux on Surfaces**

One of the crucial factors in determining how long a surface can be maintained clean (or, alternatively, how long it takes to build-up a certain surface concentration of adsorbed species) is the number of gas molecules impacting on the surface from the gas phase.

The incident *flux* is the number of incident molecules per unit time per unit area of surface.

For a given set of conditions ($P,T$ etc.) the flux is readily calculated using a combination of the ideas of statistical physics, the ideal gas equation and the Maxwell-Boltzmann gas velocity distribution.

*Step 1*: it can be readily shown that the incident flux, $F$, is related to the gas density above the surface by

$$F = \frac{1}{4} n\bar{C} \qquad \text{[molecules m}^{-2}\text{ s}^{-1}\text{]}$$

where: $n$ - molecular gas density [molecules $m^{-3}$]

$c$ - average molecular speed [m $s^{-1}$]

*Step* 2: the molecular gas density is given by the ideal gas equation, namely

$n = N / V = P / (kT)$ [molecules $m^{-3}$]

*Step* 3: the mean molecular speed is obtained from the Maxwell-Boltzmann distribution of gas velocities by integration, yielding

$$\bar{c} = \sqrt{\frac{8kT}{m\pi}} \quad [ms^{-1}]$$

where: $m$ – molecular mass [kg]

$k$ – Boltzmann constant (= $1.38 \times 10^{-23}$ $JK^{-1}$)

$T$ – temperature [K]

$\pi = 3.1416$

Step 4: combining the equations from the first three steps gives the Hertz-Knudsen formula for the incident flux $F = \frac{P}{\sqrt{2\pi mkT}}$

[molecules $m^{-2}s^{-1}$]

**Gas Exposure - the "Langmuir"**

The *gas exposure* is measure of the amount of gas which a surface has been subjected to. It is numerically quantified by taking the product of the pressure of the gas above the surface and the time of exposure (if the pressure is constant, or more generally by calculating the integral of pressure over the period of time of concern).

Although the exposure may be given in the SI units of Pa s (Pascal seconds), the normal and far more convenient unit for exposure is the *Langmuir*, where 1 L = $10^{-6}$ Torr s. i.e.

(Exposure/L) = $10^6 \times$ (Pressure/Torr) $\times$ (Time/s)

**Sticking Coefficient & Surface Coverage**

The *sticking coefficient, S,* is a measure of the fraction of incident molecules which adsorb upon the surface i.e. it is a probability and lies in the range 0 - 1, where the limits correspond to no adsorption and complete adsorption of all incident molecules respectively. In general, *S* depends upon many variables i.e.

$$S = f\,(\text{surface coverage, temperature, crystal face....})$$

The *surface coverage* of an adsorbed species may itself, however, be specified in a number of ways:

1. As the number of adsorbed species per unit area of surface (e.g. in molecules $cm^{-2}$).
2. As a fraction of the maximum attainable surface coverage i.e.

$$\theta = \frac{\text{Actual surface coverage}}{\text{Saturation surface coverage}}$$

- in which case q lies in the range 0 – 1.

3. Relative to the atom density in the topmost atomic layer of the substrate i.e.

$$q = \frac{\text{No.of adsorbed species per unit area of surface}}{\text{No.of surface substrate atoms per unit area}}$$

in which case $\theta_{max}$ is usually less than one, but can for an adsorbate such as H occasionally exceed one.

1. Whichever definition is used, the surface coverage is normally denoted by the Greek q.
2. The second means of specifying the surface coverage is only usually employed for adsorption isotherms (e.g. the Langmuir isotherm). The third method is the most generally accepted way of defining the surface coverage.
3. A *monolayer* (1 ML) of adsorbate is taken to correspond to the maximum attainable surface concentration of adsorbed species bound to the substrate.

This is dependent upon the flux of gas phase molecules incident upon the surface, the actual coverage corresponding to the monolayer and the coverage-dependent sticking probability... however, it is possible to get a minimum estimate of the time required by assuming a unit sticking probability (i.e. $S = 1$) and noting that monolayer coverages are generally of the order of $10^{15}$ per cm$^2$ or $10^{19}$ per m$^2$. Then

Time / ML ~ ($10^{19}$ / $F$)

### Variation of Parameters with Pressure

All values given below are approximate and are generally dependent on factors such as temperature and molecular mass.

| Degree of Vacuum | Pressure (Torr) | Gas Density (molecules m-3) | Mean Free Path (m) |
|---|---|---|---|
| Atmospheric | 760 | 2 × 1025 | $7 \times 10^{-8}$ |
| Low | 1 | 3 × 1022 | $5 \times 10^{-5}$ |
| Medium | $10^{-3}$ | 3 × 1019 | $5 \times 10^{-2}$ |
| High | $10^{-6}$ | 3 × 1016 | 50 |
| UltraHigh | $10^{-10}$ | 3 × 1012 | 5 × 105 |

We can therefore conclude that the following requirements exist for:

Collision Free Conditions => $P < 10^{-4}$ Torr

Maintenance of a Clean Surface => $P < 10^{-9}$ Torr

For most surface science experiments there are a number of factors necessitating a high vacuum environment:

1. For surface spectroscopy, the mean free path of probe and detected particles (ions, atoms, electrons) in the vacuum environment must be significantly greater than the dimensions of the apparatus in order that these particles may travel to the surface and from the surface to detector without undergoing any interaction with residual gas phase molecules. This requires pressures better than $10^{-4}$ Torr. There are, however, some techniques, such as IR spectroscopy, which are "photon-in/photon-out" techniques and do not suffer from this requirement.

2. (On a practical level, it is also the case that the lifetime of channeltron and multiplier detectors used to detect charged particles is substantially reduced by operation at pressures above $10^{-6}$ Torr).
3. Most spectroscopic techniques are also capable of detecting molecules in the gas phase; in these cases it is preferable that the number of species present on the surface substantially exceeds those present in the gas phase immediately above the surface - to achieve a surface/gas phase discrimination of better than 10:1 when analysing ca. 1% of a monolayer on a flat surface this requires that the gas phase concentration is less than ca. $10^{12}$ molecules $cm^{-3}$ (= $10^{18}$ molecules $m^{-3}$), i.e. that the (partial) pressure is of the order of $10^{-4}$ Torr or lower.
4. In order to begin experiments with a reproducibly clean surface, and to ensure that significant contamination by background gases does not occur during an experiment, the background pressure must be such that the time required for contaminant build-up is substantially greater than that required to conduct the experiment i.e. of the order of hours. The implication with regard to the required pressure depends upon the nature of the surface, but for the more reactive surfaces this necessitates the use of UHV (i.e. $< 1 \times 10^{-9}$ Torr).

It is clear therefore that it is the last factor that usually determines the need for a very good vacuum in order to carry out reliable surface science experiments.

## THE PRESSURE DEPENDENCE OF GAS PHASE AND SURFACE ADSORPTION CHARACTERISTICS

A limited number of examples of the application of the formulae given in the previous section to determine the:

- Density of Molecules in the Gas Phase
- Mean Free Path of Molecules in the Gas Phase
- Flux of Molecules incident upon a Surface

- Rate of Adsorption of Molecules and Surface Coverages

Within any one of the following sub-sections, it will be assumed that you have already done the previous questions and may make use of the answers from these questions - you are therefore advised to work through the questions in the order they are presented.

**Molecular Gas Densities**

What is the molecular gas density for an ideal gas at 300 K when the pressure is $10^{-6}$ Torr ? (in molecules $m^{-3}$)

What therefore would be the gas density at a pressure of $10^{-9}$ Torr ?

**Mean Free Path of Molecules in the Gas Phase**

Calculate the mean free path of CO molecules in a vessel at a pressure of $10^{-4}$ Torr at 300 K. (Collision cross section = 0.42 $nm^2$).

Now calculate the mean free path of CO molecules, again at 300 K, at a pressure of $10^{-9}$ Torr.

**Fluxes of Molecules Incident upon a Surface**

Calculate the flux of molecules incident upon a solid surface under the following conditions:

[1 u = $1.66 \times 10^{-27}$ kg: atomic masses; $m$(O) =16 u, $m$(H) = 1 u]

Oxygen gas (P = 1 Torr) at 300 K

Oxygen gas (P = $10^{-6}$ Torr) at 300 K

Hydrogen gas (P = $10^{-6}$ Torr) at 300 K

Hydrogen gas (P = $10^{-6}$ Torr) at 1000 K

**The Kinetically Limited Uptake of Molecules onto a Surface**

The rate of adsorption of molecules onto a surface can be determined from the flux of molecules incident on the surface and the sticking probability pertaining.

# 10

# Synthesis and Assembly of Materials

The corresponding means of synthesis and assembly are similarly wide-ranging. But however multifaceted the synthesis approaches and the ultimate applications, there are common issues and unique defining features of these nanostructured materials.

First, there is the recognition of critical scale lengths that define the material structure and organization, generally in the nanometer range; and that ultimately determine the fundamental macroscopic properties of the material. Research in nanostructured materials is motivated by the belief that ability to control the building blocks or nanostructure of the materials can result in enhanced properties at the macroscale: increased hardness, ductility, magnetic coupling, catalytic enhancement, selective absorption, or higher efficiency electronic or optical behaviour.

Synthesis and assembly strategies accommodate precursors from liquid, solid, or gas phase; employ chemical or physical deposition approaches; and similarly rely on either chemical reactivity or physical compaction to integrate nanostructure building blocks within the final material structure. The variety of techniques is shown schematically in Figure.

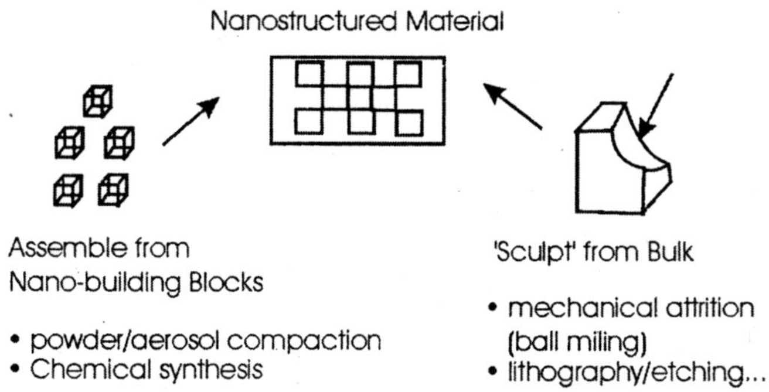


**Fig.** Schematic of variety of nanostructure synthesis and assembly approaches.

The "bottom-up" approach first forms the nanostructured building blocks and then assembles them into the final material. An example of this approach is the formation of powder components through aerosol techniques and then the compaction of the components into the final material. These techniques have been used extensively in the formation of structural composite materials. One "top-down" approach begins with a suitable starting material and then "sculpts" the functionality from the material. This technique is similar to the approach used by the semiconductor industry in forming devices out of an electronic substrate (silicon), utilizing pattern formation (such as electron beam lithography) and pattern transfer processes (such as reactive ion etching) that have the requisite spatial resolution to achieve creation of structures at the nanoscale. This particular area of nanostructure formation has tremendous scope, warranting its own separate study, is a driving issue for the electronics industry, and will not be a principal theme of this study. Another top-down approach is "ball-milling," the formation of nanostructure building blocks through controlled, mechanical attrition of the bulk starting material. Those nano building blocks are then subsequently assembled into a new bulk material.

In fact, many current strategies for material synthesis integrate both synthesis and assembly into a single process, such as characterizes chemical synthesis of nanostructured materials. The degree of control required over the sizes of the nanostructure components, and the nature of their distribution and bonding within the fully formed material varies greatly, depending on the ultimate materials application. Achieving selective optical absorption in a material (e.g., UV-blocking dispersions) may allow a wide range of sizes of the component nanostructure building blocks, while quantum dot lasers or single electron transistors require a far tighter distribution of size of the nanostructure components. Compaction methods may provide excellent adhesion for nanocomposite materials of improved structural performance (e.g., ductility), but such interfaces may be unsatisfactory for electronic materials.

## CRITICAL ISSUES FOR NANOSTRUCTURE SYNTHESIS AND ASSEMBLY

However broad the range of synthesis approaches, the critical control points fall into two categories:

1. Control of the size and composition of the nanocluster components, whether they are aerosol particles, powders, semiconductor quantum dots, or other nanocomponents
2. Control of the interfaces and distributions of the nanocomponents within the fully formed materials

These two aspects of nanostructure formation are inextricably linked; nevertheless, it is important to understand how to exercise separate control over the nucleation of the nanostructure building blocks and the growth (for example, minimizing coagulation or agglomeration) of those components throughout the synthesis and assembly process. This latter issue is related to the importance of the following:

- The chemical, thermal, and temporal stability of such formed nanostructures
- The ability to scale-up synthesis and assembly strategies for low-cost, large-scale production of nanostructured materials, while at the same time maintaining control of critical feature size and quality of interfaces (economic viability is a compelling issue for any nanostructure technology)

All researchers in this area are addressing these issues.

## COMMON ENABLING TECHNOLOGIES

There has been steady technological progress in all fields of nanostructure synthesis and assembly, in no small part because of the more general availability of characterization tools having higher spatial, energy, and time resolution to clearly distinguish and trace the process of nanostructure formation. As transmission electron microscopy and X-ray diffraction techniques helped in an earlier period to relate the improved properties of "age-hardened" aluminum alloys to their nanostructure, today's technological advances in materials characterization are providing new insights into the role of the nanostructure in determining macroscopic properties. The tightly-coupled iteration between characterizing the nanostructure, understanding the relationship between nanostructure and macroscopic material properties, and improved sophistication and control in determining nanostructure size and placement have accelerated the rate of progress and helped to define the critical components of this "new" field of nanostructure science and technology. Tightly focused (1-2 μm), high brightness synchrotron X-ray sources provide detailed structural information on colloids, polymers, alloys, and other material structures, highlighting the inhomogeneities of the material with suitable spatial resolution.

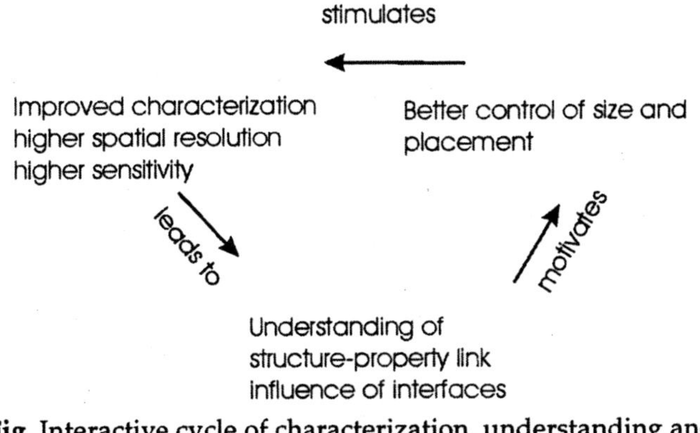


**Fig.** Interactive cycle of characterization, understanding and enhanced control in the synthesis and assembly of nanostructures.

Another important enabling technology has been the now widely available scanning probe technology, including scanning tunneling microscopy and atomic force microscopy. The power of these techniques has provided impetus for developing even higher performance scanning probe tips, fabricated through microfabrication techniques. Development of different tip structures in various materials has given rise to an entire family of powerful scanning probe techniques that encompass such a wide range of characterization capabilities that one can envision "a laboratory on a tip". The development of a tip technology also impacts the synthesis and assembly processes themselves: scanning probe technologies have been used as the basis of materials patterning and processing at nanometer scales and have provided information on the mechanical and thermal properties of materials at the nanoscale.

More sophisticated in situ monitoring strategies have provided greater understanding and control in the synthesis of nanostructured building blocks, particularly those formed in vacuum environments. Molecular beam epitaxy (MBE) represents a physical vapour (gas phase) deposition technique where sub-monolayer control can be imposed on the formation of two-dimensional and, more recently, three-dimensional nanostructured materials. A great deal of the understanding and control derives from the ability to carry out sensitive monitoring of the growth process in situ: reflection high energy electron diffraction (RHEED) details the nature of the surface and surface bonding, and oscillations of the RHEED intensity provide information on the growth rate.

The improvements brought about by these advances in technology have been substantial, but perhaps of greater importance for this nascent field of nanostructure science and technology has been the development of strategies and technologies that have been formed across the former disciplines. More reliable means of controlling nanostructure size and placement, with an end view of being able to scale up the production of such materials while maintaining the control over the nanostructure, have given impetus to a common search for novel synthesis and assembly strategies. In that search, it is apparent that the naturally occurring synthesis and assembly of biological materials can provide us with some critical insights.

## NANOPARTICLE SYNTHESIS STRATAGIES

### Gas Phase Synthesis and Sol-Gel Processing

Major efforts in nanoparticle synthesis can be grouped into two broad areas: gas phase synthesis and sol-gel processing. Nanoparticles with diameters ranging from 1 to 10 nm with consistent crystal structure, surface derivatization, and a high degree of monodispersity have been processed by both gas-phase and sol-gel techniques. Typical size variances are about 20%; however, for measurable enhancement of the quantum effect, this must be reduced to less than 5%.

Initial development of new crystalline materials was based on nanoparticles generated by evapouration and condensation (nucleation and growth) in a subatmospheric inert-gas environment. Various aerosol processing techniques have been reported to improve the production yield of nanoparticles. These include synthesis by combustion flame, plasma, laser ablation, chemical vapour condensation, spray pyrolysis, electrospray, and plasma spray.

Sol-gel processing is a wet chemical synthesis approach that can be used to generate nanoparticles by gelation, precipitation, and hydrothermal treatment. Size distribution of semiconductor, metal, and metal oxide nanoparticles can be manipulated by either dopant introduction or heat treatment. Better size and stability control of quantum-confined semiconductor nanoparticles can be achieved through the use of inverted micelles, polymer matrix architecture based on block copolymers or polymer blends, porous glasses, and ex-situ particle-capping techniques.

## OTHER STRATEGIES

Additional nanoparticle synthesis techniques include sonochemical processing, cavitation processing, microemulsion processing, and high-energy ball milling. In sonochemistry, an acoustic cavitation process can generate a transient localized hot zone with extremely high temperature gradient and pressure. Such sudden changes in temperature and pressure assist the destruction of the sonochemical precursor (organometallic solution) and the formation of nanoparticles. The technique can be used to produce a large volume of material for industrial applications.

In hydrodynamic cavitation, nanoparticles are generated through creation and release of gas bubbles inside the sol-gel solution. By rapidly pressurizing in a supercritical drying chamber and exposing to cavitational disturbance and high temperature heating, the sol-gel solution is mixed. The erupted hydrodynamic bubbles are responsible for nucleation, growth, and quenching of the nanoparticles. Particle size can be controlled by adjusting the pressure and the solution retention time in the cavitation chamber.

Microemulsions have been used for synthesis of metallic, semiconductor, silica, barium sulfate, magnetic, and superconductor nanoparticles. By controlling the very low interfacial tension ($\sim 10^{-3}$ mN/m) through the addition of a cosurfactant (an alcohol of intermediate chain length), these microemulsions are produced spontaneously without the need for significant mechanical agitation. The technique is useful for large-scale production of nanoparticles using relatively simple and inexpensive hardware.

Finally, high energy ball milling, the only top-down approach for nanoparticle synthesis, has been used for the generation of magnetic, catalytic, and structural nanoparticles. The technique, which is already a commercial technology, has been considered dirty because of contamination problems from ball-milling processes. However, the availability of tungsten carbide components and the use of inert atmosphere and/or high vacuum processes have reduced impurities to acceptable levels for many industrial applications. Common drawbacks include the low surface area, the highly polydisperse size distributions, and the partially amorphous state of the as-prepared powders.

## MEANS TO ACHIEVE MONODISPERSITY

One of the most challenging problems in synthesis is the controlled generation of monodispersed nanoparticles with size variance so small that size selection by centrifugal precipitation or mobility classification is not necessary. Gas-phase synthesis is one of the best techniques with respect to size monodispersity, typically achieved by using a combination of rigorous control of nucleation-condensation growth and avoidance of coagulation by diffusion and turbulence as well as by the effective collection of nanoparticles and their handling afterwards. The stability of the collected nanoparticle powders against agglomeration, sintering, and compositional changes can be ensured by collecting the nanoparticles in liquid suspension. For semiconducting particles, stabilization of the liquid suspension has been demonstrated by the addition of polar solvent, surfactant molecules have been used to stabilize the liquid suspension of metallic nanoparticles. Alternatively, inert silica encapsulation of nanoparticles by gas-phase reaction and by oxidation in colloidal solution has been shown to be effective for metallic nanoparticles.

New approaches need to be developed for the generation of monodisperse nanoparticles that do not require the use of a size classification procedure. An example of this is a process developed in Japan where very monodispersed gold colloidal nanoparticles with diameters of about 1 nm have been prepared by reduction of metallic salt with UV irradiation in the presence of dendrimers. Poly(amidoamine) dendrimers with surface amino groups of higher generations have spherical 3-D structures, which may have an effective protective action for the formation of gold nanoparticles. Although the specific role of dendrimers for the formation of monodispersed nanoparticles has yet to be defined, good monodispersity is thought to come from the complex reaction accompanying the decomposition of dendrimers, which eventually leads to the conversion of solution ions to gold nanoparticles.

## SCALEUP

Scaleup production is of great interest for nanoparticle synthesis. High energy ball milling, already a commercial high-volume process, as mentioned above, has been instrumental in generating nanoparticles for the preparation of magnetic, structural, and catalytic materials. However, the process produces polydispersed amorphous powder, which requires subsequent partial recrystallization before the powder is consolidated into nanostructured materials. Although gas-phase synthesis is generally a low production rate process (typically in the 100 milligrams per hour range) in research laboratories, higher rates of production (about 20 grams per hour). Even higher production rates (about 1 kg per hour) are now being achieved commercially. For sol-gel processing, the development of continuous processing techniques based on present knowledge of batch processing has yet to be addressed for economical scaleup production of nanoparticles. Other related sol-gel issues concern the cost of precursors and the recycling of solvent. Overall, sol-gel processing is attractive for commercial scale-up production.

## BUILDING NANOPARTICLE CHAINS OR WIRES

A recent paradigm shift envisioned for optoelectronics and computational devices involves the assembly of molecular or quantum wires. Large polymeric molecules have been used as nano building blocks for nanoporous molecular sieves, biocompatible materials, optical switching, data processing, and other nonlinear optical components. Chain aggregates of nanoparticles can be considered as polymer-like units with their primary particles composed of a few hundred to a few thousand molecules. Thus, these chain aggregates can be considered "heavy" quantum wires. In fact, nanoparticle chain aggregates have been studied extensively as magnetic materials, as reinforced elastomers, and as additives in concrete. These aggregates have been shown to have chemical and mechanical properties different from those of individual primary particles. Depending on the particle size and its compositional material, the bonding force responsible for holding the aggregates together varies from weak van der Waals force for micrometer particles to strong chemical bonds for nanometer particles to very strong magnetic dipolar bonds for nanosized magnetic particles. The mechanical, optical, and electronic transport properties can be varied by controlling the diameter and the monodispersity of the primary particles, the crystalline structure and morphology, aggregate length, interfacial properties, and material purity. These chain aggregates can be formed by allowing agglomeration of nanoparticles generated by any of the synthesis techniques, with the exception of high energy ball milling, which generates particles with low surface area and high anisotropic morphologies, both of which are detrimental for the formation of chain aggregates. Depending on the magnetic and electric charging properties of the nanoparticles, an external applied magnetic or electric field can be used to control the fractal dimension of aggregates. For optical applications of chain aggregates, lower fractal dimensions (i.e., relatively straight chain aggregates with few branches) are desirable.

## BUILDING NANOMETER FIBRES OR TUBES

Recent advances in the fabrication of nanometer fibres or tubes offer another form of building blocks for nanostructured materials. An effective way to generate nanometer fibres (or tubes) is based on the use of membrane-template techniques. Membranes, with nanochannels generated by fission-fragment tracks or by electrochemical etching of aluminum metal, are used as templates for either chemical or electrochemical deposition of conductive polymers, metal, semiconductor, and other materials for the generation of nanofibers or tubes. Since the nanochannels on membranes are very uniform in size, the diameter and the aspect ratio of the nanofibers (or tubes) synthesized by the membrane-template technique can be precisely controlled. This greatly facilitates the interpretation of optical data and the processing of these fibres (or tubes) into 2-D nanostructured materials. Single-crystal semiconductor nanofibers can also be grown catalytically by metalorganic vapour phase epitaxy and laser ablation vapour-liquid-solid techniques. The synthesis of these one-dimensional structures with diameters in the range of 3 to 15 nm holds considerable technological promise for optoelectronic device applications, such as the p-n junctions for light emission.

The advent of carbon-based nanotubes has created yet another way to fabricate nanometer fibres and tubes. These nanotubes have been used as templates for the fabrication of carbide and oxide nanotubes. Synthesis of nanotubes based on BN, BC3 and BC2N have also been reported. These nanotubes potentially possess large third-order optical non-linearity and other unusual properties. Metallic nanofibers synthesized by carbon-nanotube-template techniques are useful in the design of infrared absorption materials. The carbon nanotubes can now be catalytically produced in large quantities and have been used for reinforcement of nanostructural composite materials and concrete.

## BIOGENIC STRATEGIES

The elegant complexity of biological materials represents the achievement of structural order over many length scales, with the full structure developed from the "nested levels of structural hierarchy", in which self-assembled organic materials can form templates or scaffolding for inorganic components. These notions of a multilevel material structure with strong interactions among levels and an interplay of perfection and imperfection forming the final material. Along with characteristic length scales, there are characteristic relaxation times of the material, bringing in the consideration of the temporal stability of the structured materials.

## SELF-ASSEMBLY AS A DELIBERATE STRATEGY

"Self-assembly" is a term that by now figures prominently in the literature of nanostructured materials and nanofabrication. The term therefore carries a variety of implicit or explicit meanings, and we cite the definition given by Kuhn and Ulman: "Self-assembly is a process in which supermolecular hierarchical organization is established in a complex system of interlocking components." The mechanisms that produce the hierarchical organization are determined by competing molecular interactions (interactions between hydrophobic versus hydrophilic components, van der Waals, Coulombic, or hydrogen bonding), resulting in particular microphase separation or surface segregation of the component materials. Thus, the use of a hierarchy of bond strengths and/ or chemical specificity can produce a hierarchy of lengths in the final nanostructured material.

As one example of such self-assembled or self-organized materials mixed silica precursors with surfactants that have self-ordered to form various surfactant-water liquid crystals, producing various structures built from walls of amorphous silica, organized about a repetitive arrangement of pores up to a hundred angstroms in diameter. A range of such structures. In the "natural" formation of inorganic nanostructures, the addition of organic molecules can strongly influence the resulting structure of inorganic components. Such strategies have been adopted in synthetic formation of nanostructures, such as in the formation of networks of gold clusters.

Gold clusters 3.7 nm in diameter, formed in the vapour phase, are encapsulated in organic surfactants, such as dodecanethiol, forming a colloidal suspension. The surfactants prevent the agglomeration of the gold clusters. Addition of small amount of dithiol precipitates out a 3-D cluster network, which can in turn be deposited onto another solid substrate. Figure shows a transmission electron microscope (TEM) image of a cluster array spin-cast onto $MoS_2$.

The methodology of self-assembly has even been extended to physical vapour deposition processes where it would seem more difficult to control the nucleation and growth of three-dimensional nanostructures. Utilizing the strain inherent in the epitaxial growth of lattice-mismatched materials, and the expected strain-induced transition from two-dimensional (layered) to three-dimensional (islanded) growth, together with careful monitoring of the growth process through RHEED analysis, researchers have been able to form arrays of semiconductor quantum dots, ~ 200-300 Å in diameter, ~ 1011 $cm^{-2}$ in density, and with a size variation of about ±7%. An example of such "self-assembled" semiconductor quantum dots. The achievement of arrays of several billions of quantum dots of these dimensions with such a size variation is beyond the capability of standard high resolution lithographic and pattern transfer processes. Moreover, the controlled formation of critical surfaces and interfaces without the intercession of ion-assisted processing that can introduce potential defects into the materials has produced a rich source of optically and electronically efficient quantum structures. A number of researchers have already incorporated such self-assembled dots into laser structures.

Chemical specificity may provide the most robust means of ensuring control of size and placement of nanostructured building blocks, and recent work in the synthesis of compound semiconductor quantum dots from chemical precursors have provided even tighter distributions of size variation (±5%) than those shown in the strain-induced self-assembled dots. The size-dependent and energy-selective optical absorption properties of such chemically synthesized quantum dots.

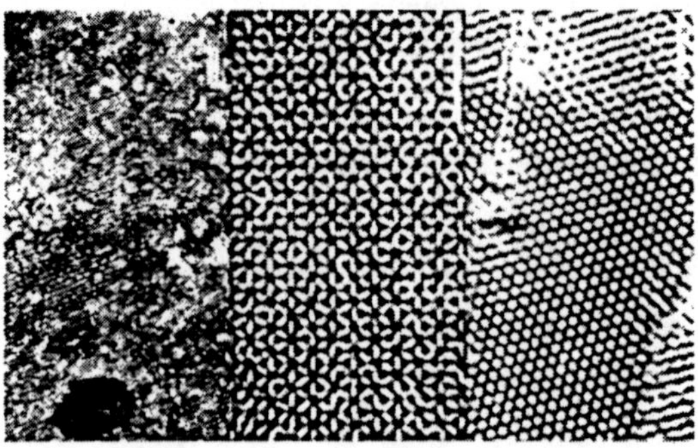

**Fig:** TEM images of (a) the lamellar morphology, (b) the cubic phase with Ia3d symmetry viewed along its [111] zone axis, and (c) the hexagonal phase viewed along its [001] zone axis of thesilica/ surfactant nanostructured composites by co-assembly (bars = 30 nm).

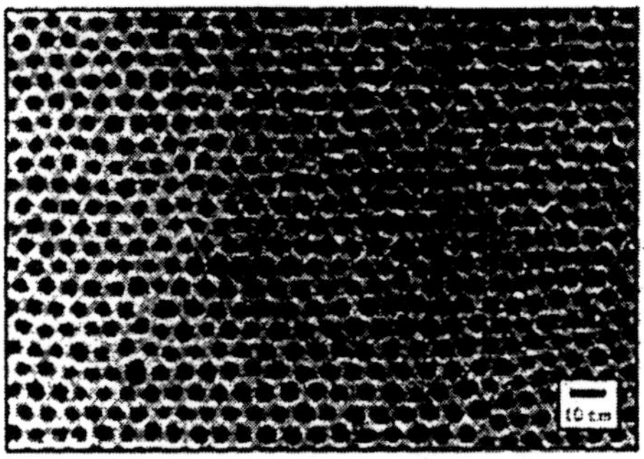


**Fig:** TEM image of unlinked cluster array of 3.7 nm Au clusters encapsulated by dodecanethiol.

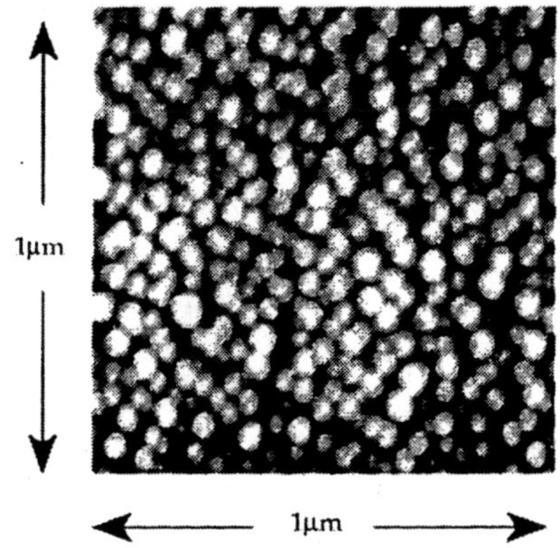


**Fig:** Array of InAs quantum dot structures grown on GaAs substrates.

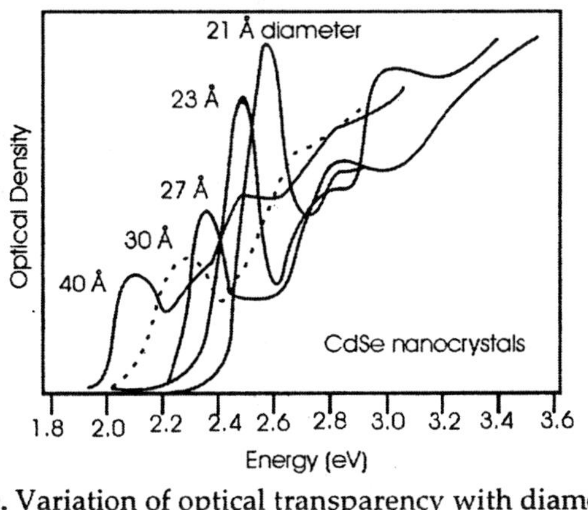


**Figure.** Variation of optical transparency with diameter of chemically synthesized CdSe nanocrystals.

## NATURAL TEMPLATES OR "SCAFFOLDS"

Biogenic systems employ natural templates or scaffolding in the construction of nanoscaled materials, where the templates and scaffolding can help set the proper, critical dimensions that are characteristic of the final material. The templates can be formed "artificially," through the use of lithographically defined patterning and processing of the substrate material in order to achieve selective nucleation and growth of the nanostructured materials. However, there are also a host of "natural" templating materials that researchers can avail themselves of. The zeolites have for a long time been utilized as the basis of high surface area materials that enable catalysis. Recent work has taken these natural cage structures and formed large cage structures with varying pore sizes. When zeolite cages are "loaded" with various materials, the controlled proximity enforced by the cage structure can result in various magnetically active or nonlinear optical materials.

More recently, the C-based fullerenes and C-nanotubes have provided a tremendously rich basis for nanostructure research;

The by now "easy" formation of such structures and ready availability of such materials provide a rich source of nanostructures with nearly immediate applications for electronic devices, storage, and enhancement of structural materials. Further, these materials can serve as the nanostructured precursors of nanostructure synthesis approaches (such as the starting material for cluster or aerosol deposition), or can provide an avenue for the synthesis of templates of varying sizes. Such templates can be electronically "doped" and filled with other materials. Recent work reported the synthesis of "nanorods" of GaN through confined reactions in carbon nanotubes.

## DISPERSIONS AND COATINGS

## INTRODUCTION

The object of the controlled assembly of nanoparticulates is to make materials with new properties and assemble them with practical applications. A unique value of nanoparticulates is their extremely high particle surface area; having many more sites for achieving property enhancements makes them ideal for a wide variety of applications as dispersions and coatings. Dispersive and coating applications of nanoparticles include optical, thermal, and diffusion barriers. Significant work on nanoscale dispersions and coatings is underway worldwide in the areas of ceramics, cosmetics, biosensors, colourants, and abrasion-resistant polymers. Other applications include imaging ink jet materials, electrophotography, pharmaceuticals, flavour enhancers, pesticides, lubricants, and other proprietary applications specific to industry. Still another application is in a new, post-silicon generation of electronic devices that includes nanotubes and fullerenes as constituent units of carbon nanoelectronic devices; here, dispersion takes on a more quantum consideration in which the number of atoms in a cluster is compared to the number of surface atoms to determine its dispersion function. Also in the semiconductor industry, a monolayer or thin film coating of atoms or molecules is deposited on foils, metal sheets, or glass to enhance storage capacity and accelerate responses from the electronic component.

All these applications deal with dispersions or coatings of particles that enhance specific features. Also, ability to manufacture smaller functional systems enhances performance, cost, and efficiency. These considerations power the drive towards miniaturization and precision finishing.

## PROPERTY AMPLIFICATION

What's particularly exciting in the study of nanoparticles, nanostructured materials, and nanodevices is their ability to also add value to materials and products through enhancement of specific properties, such as the following:

- Mechanical strength. Nanostructured powders have been produced by plasma processing where the reactor vapourizes coarse metal particles; by combustion synthesis where redox reactions take place at elevated temperatures, followed by quenching; and by mechanical alloying with gas atomization. When nanocrystalline powders achieved by these means are compacted and applied as a coating, they lend significant strength and ductility to a variety of conventional materials such as ceramic, composites, and metal alloys.

- Superconductivity. The nature of the superconductivity effect in detail. It involves a deposition of nanocrystals on substrates, leading to improved optical and electrical properties.
- Covering power. Because nanostructuring increases the number of active sites-there are many more atoms per grain boundary-the enhanced surface area leads to a reduced material requirement, which in turn can lower cost.
- Ability to incorporate high cost materials. Expensive materials such as colourants and drugs may be effectively dispersed in small and controlled quantities through nanostructuring.
- Environmental value. Improvements in environmental impact are achieved by utilizing nanostructure particulates in coatings and thus eliminating the requirement for toxic solvents. By eliminating hazardous wastes, nanocoatings can both reduce a company's disposal costs and improve its environmental position.

Thus, nanostructured dispersions and coatings can significantly reduce material costs and improve performance and functionality in a large variety of applications. In all of the sites the WTEC panel visited in Europe, Russia, and Japan, research groups were interested in achieving one or more of the properties listed above.

## ENABLERS

For nanoparticulate dispersions and coatings, certain enablers must be present to achieve success:

1. Effective particle preparation,
2. Stabilization of the dispersed phase,
3. Scaleup and control of the process,
4. The existence of excellent analytical capabilities. The main methods or issues for each are described below.
    a. Particle preparation. Wet chemical methods such as liquid phase precipitation or sol-gel methods are of high interest in particle preparation. In the area of hybrid methods, both spray pyrolysis and flame hydrolysis are utilized. Numerous physical methods such as mechanical size reduction are also often employed.
    b. Stabilization of the dispersed phase. For stabilization of the dispersed phase, it is necessary to understand how particles can be kept as distinct entities. Either a charged stabilized system or a sterically stabilized approach is required. A successful preparation for use as a dispersion should be free from agglomeration in the liquid state so as to maintain particle integrity. The dry-coated format of nanoparticles should also minimize any presence of particle aggregation
    c. Scaleup and control of the process. Here, the issue in achieving a high rate of production of powder is the effective pyrolysis of the gas stream containing the precursor. Issues such as scaling and reproducibility from one run of nanoparticle material or dispersions to another are all inherent in the concept of precision engineering or invariant process control.
    d. Analytical capabilities. Analytical capabilities are absolutely essential for characterizing dispersions and coatings. Particle size determinations, assay analysis, and interfacial properties are all important. Transmission electron microscopes, atomic force microscopy, nuclear magnetic resonance, and scanning tunneling microscopy are

just some of the tools utilized in characterizing nanoparticle dispersions and coatings, particularly at the very small end of the nanoscale.

## APPLICATIONS

Below are some general examples of the vast array of applications of nanoparticulate dispersions and coatings. There are a number of applications in all of the categories below that are already in the public domain; however, most are still highly proprietary, especially drug delivery applications.

Cosmetics. An area of nanoparticle technology that has incredible commercial potential is the cosmetic industry. Here there is a great demonstrated demand, and the technology can be made simple, since properties of colour and light fastness are achieved by component mixing in the cosmetic preparation.

In the area of medical applications, finely dispersed pharmaceuticals offer rapid drug delivery and reduced dosages for patients. Dispersions of strong and resilient biocompatible materials suggest opportunities for artificial joints. These generally are ceramic materials containing nanoparticulates.

Overall, much of the demand for nanoparticulate dispersions and coatings comes from the cosmetic and pharmaceutical industries; in particular, liquid dispersion preparations will be widely used to apply topical coatings to the human epidermis because they can be absorbed faster and more completely than conventional coatings.

Microelectromechanical systems (MEMS). Although MEMS technologies will support the semiconductor industry in particular, there are many other applications being explored, such as in medicine, ceramics, thin films, metal alloys, and other proprietary applications. In the United States a particular focus is applying sputtering coatings to achieve MEMS technology in concert with these applications.

Printing. In the areas of image capture/image output addressed by ink jet technology, nanoscience can help control the properties of the inks themselves. The production and use of nanoengineered ink products benefits from such complimentary technology as laser-assist delivery of the ink jet droplet to maintain an accurate deposit of the ink on its target. Another application in this field is using nanoscale properties to tailor the inks to achieve ideal absorption and drying times for desired colour properties and permanency.

Semiconductors. One form of "bottom up" technology that is receiving considerable attention is thin films for the semiconductor industry. Here single atoms or molecules are deposited by physical vapour deposition, which could be achieved through sputtering, molecular beam epitaxy, or chemical vapour deposition. Sputtering is used on a large scale to coat metal sheets, glass, polymer substrates and other receptive materials in order to produce enhanced electronic properties for information storage and processing speed.

Sensors. Chemical or physical sensors often use nanoparticles because they provide high surface area for detecting the state of chemical reactions, because the quality of detection signals is improved, and because earlier and more accurate determination of leakage reduces waste. Some commercial sensors and actuators composed of thin films are already used for environmental vapour monitoring in reactors, for example.

Other likely applications of nanotechnology involving dispersions and coatings include nanofabricated surface coatings for keeping windows and surfaces clean. Here, the transparent nanocoating on a surface prevents fog and dirt particles from depositing on the substrate. Commercial products also include aluminum oxide/epoxy dispersions yielding 19 times more wear resistance than conventional products.

The WTEC panelists were privileged to hear of and to observe a wide variety of current work on nanostructured dispersions and coatings in various laboratories around the world. Descriptions follow of a number of research projects and the preparation issues being addressed in the United States and in some of the foreign laboratories that panelists visited. Table outlines some of the nanoparticle preparation techniques that are currently in use.

Some of the highlights of that volume are summarized below. Similar work is ongoing in other countries.

**Table. Nanoparticle Preparation for Dispersions/Coatings**

| Liquid Chemical Methods | Physical Methods | Elevated Temperature Methods |
|---|---|---|
| Sol-Gel * | Size reduction by mechanical means (low energy) | Aerosol |
| Chemical Precipitation | | Flame |
| Ag-X (Eastman Kodak) † | | |
| Colloidal Micelles | | Vapour phase condensation |

A ceramic thin film structure that would mirror-image a self-assembly process in many materials. Here silica precursors when mixed with surfactants yield polymerized templates having structures similar to surfactant-water liquid crystals; what results are highly controlled pores on the 10 to 100 Å scale. Controlling the pore structure and synthesizing the building blocks are two technical challenges facing future work in this area. However, Aksay has shown that he can grow silicate films onto a wide variety of substrates. The layer structure of the bound surfactant molecules is key. Atomic force microscopy can reveal some details on the morphology of these films. The nanostructural patterns obtained in processing ceramics that contain organic/inorganic composites allow self-assembly to take place at lower temperatures.

Another example of thin films or coating is the work of Gell, focused on improving both the physical and the mechanical properties of materials. Nanostructured coating can lead to high diffusivity, improved toughness and strength, and better thermal expansion coefficients, with lower density, elastic modulus, and thermal conductivity. In comparisons of nanostructures and conventional materials, use of nanostructured WC/cobalt composites have resulted in as much as a two-fold increase in abrasion resistance and hardness. Deposition is often accomplished by utilizing a sputtering chamber where the nanomaterial is coated on substrates. This can lead to such benefits as resistance to oxidation and cracks in addition to resistance to wear and erosion.

An area that offers exciting possibilities in the area of dispersions is the sol-gel process described by P. Wiltzius. Here, a concentrated dispersion of colloids is chemically converted into a gel body. Drying followed by sintering produces a ceramic or glass product. This process can create nanoparticles, fibres, film, plates,. or tubes. All processing is at low temperatures. Lucent Technologies has developed a silica casting process that is reproducible for making tubes of pure silica of one meter in length for use in manufacturing optical fibres. Technical challenges include obtaining pure starting materials, removing refractory particles that lead to breakage in the fibre drawing process, and achieving very tight dimensional tolerances. Colloidal dispersions of this type play a critical role in chemical mechanical polishing. To obtain good yield and high quality, it is necessary to achieve very tight process control.

At Nanophase Technologies Corporation describes the gas phase condensation process for synthesizing inorganic and metallic powders. The process involves production of physical vapour from elemental or reacted material followed by sudden condensation and reaction of the vapour into small nanometer particles. The condensation process is rapid and involves dilution to prevent the formation of hard agglomerates and coalescence.

Nanophase Technologies has developed a production system where forced convection flow controls the particle/gas stream and enhances particle transport from the particle growth region so as to generate more metal vapour. The convection flow helps in forming oxides and nitrides from metal crystallites. This process produces commercial quantities of nanosized inorganic powders that have a spherical shape with narrow size distributions.

Nanophase Technologies Corporation has developed a coating process that encapsulates nanoparticles with a surface treatment that ensures individual integrity of the particles in subsequent coating steps. Applications span their use in low dielectric media all the way to water and, if needed, steric stabilizers.

Nanophase also has directed efforts in the electronics and industrial catalyst areas. Key process parameters include time, temperature, and volume concentration. Most commercially produced particles have polydispersity. Important full-scale processes are flame reactors for preparing pigments and powdered materials for optical fibres. Pyrolysis reactors have long been used to prepare carbon blacks.

The first is a powder processing route involving preparations by physical or chemical means followed by condensation and sintering. Materials for cutting tools such as tungsten carbide/cobalt powders were prepared in a controllable way to about 50 nm. Liquid phase sintering completes the formation of the solid dispersion phase. However, a gap in fabrication technology remains that of controlling grain growth during liquid phase sintering.

A second approach to fabrication of bulk materials is spray forming. This procedure avoids contamination and coarsening of the dispersion and its particles when the process has a controlled atmosphere of inert gas at low pressure. In this approach there is need to establish process/product co-design where we understand cause-effect relationships between processing parameters and properties of nanophase spray/deposited materials.

**Table: Particle Properties**

| Particle Properties | $TiO_2$ | ZnO | $Fe_2O_3$ |
|---|---|---|---|
| Refractive Index | 2.40 | 2.01 | 3.01 |
| Density (gm/cc) | 3.95 | 5.61 | 5.30 |
| Molecular Weight (μ) | 80 | 81 | 160 |
| Particle Sizes | | | |

Optical Density (D) = Absorbance (A) = log 10 (Io/I = ε CI)

C = concentration

l = optical path length

ε = molar absorption coefficient/2.303

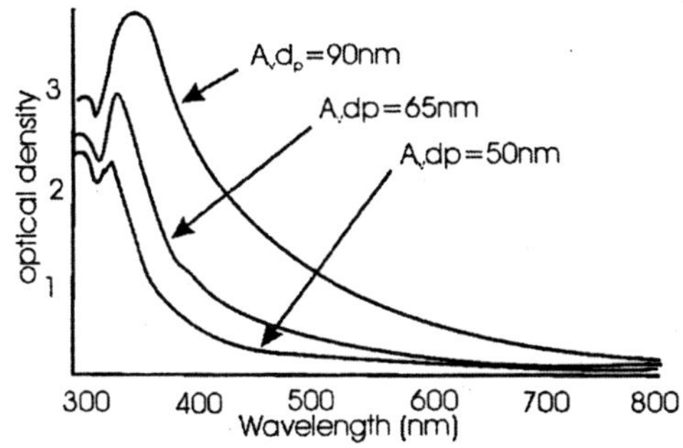


**Fig:** Transparency as a function of particle size (Nanophase).

In summary, particle preparation utilizes a wide variety of methods-both chemical and physical, at both high and low temperatures-all having unique considerations for scale-up and process control. Researchers in other countries are tackling many of these same issues.

## ALTERNATIVE PREPARATION METHODS

The quest for low temperature nanoparticle preparation methods has spanned a wide range of systems. One that has been in existence for decades but has not been put into use in other industries is the method of preparing silver halide particles. Eastman Kodak in France, England, and the United States has utilized solution precipitation technology with well-controlled mixing and nucleation control to produce a wide range of grain sizes. There are other methods of creating nanoparticles of organic materials such as filter dye applications in photographic films and spectral sensitizing dyes for use in silver halide grains. Ultrafine grinding media are used to almost sandpaper organic crystals to nanoparticle ranges of 20-80 nm. Similar technology by Bishop has been utilized in both pharmaceutical preparations and ink jet applications with good success.

One other exciting area is in polymer science, where dendrimer molecules, often 10 nanometers in diameter, are prepared synthetically. These molecules have been studied in gene therapy, as aids in helping to detect chemical or biological agents in the air, and as a means to deliver therapeutic genes in cancer cells. One can imagine applications for coatings in which these molecules with their highly reactive surfaces participate in either classical or novel nanoscopic conversions.

# Index